Sex, Drugs, Einstein, & Elves

Sushi, Psychedelics, Parallel Universes, and the Quest for Transcendence

Clifford A. Pickover

Works by Clifford A. Pickover

The Alien IQ Test
Black Holes: A Traveler's Guide
Calculus and Pizza
Chaos and Fractals
Chaos in Wonderland
Computers, Pattern, Chaos and Beauty
Computers and the Imagination
Cryptorunes: Codes and Secret Writing
Dreaming the Future
Egg Drop Soup
Future Health
Fractal Horizons: The Future Use of Fractals
Frontiers of Scientific Visualization
The Girl Who Gave Birth to Rabbits
Keys to Infinity
Liquid Earth
The Lobotomy Club
The Loom of God
The Mathematics of Oz
Mazes for the Mind: Computers and the Unexpected
Mind-Bending Visual Puzzles (calendars and card sets)
The Paradox of God and the Science of Omniscience
A Passion for Mathematics
The Pattern Book: Fractals, Art, and Nature
The Science of Aliens
Spider Legs (with Piers Anthony)
Spiral Symmetry (with Istvan Hargittai)
Strange Brains and Genius
Sushi Never Sleeps
The Stars of Heaven
Surfing Through Hyperspace
Time: A Traveler's Guide
Visions of the Future
Visualizing Biological Information
Wonders of Numbers
The Zen of Magic Squares, Circles, and Stars

Sex, Drugs, Einstein, & Elves

Sushi, Psychedelics, Parallel Universes, and the Quest for Transcendence

Clifford A. Pickover

SmartPublications™

Petaluma, California

SEX, DRUGS, EINSTEIN, & ELVES

By Clifford A. Pickover

Published by:

PO Box 4667
Petaluma, CA 94955
www.smart-publications.com

Published in the United States of America
Second Edition, 2005
Third Printing 2006

Library of Congress Control Number: 2004117266

ISBN: 1-890572-17-9   **Softcover**

Warning—Disclaimer

Praise from the Publishing Community
for *Sex, Drugs, Einstein, & Elves*

The following uplifting blurbs came from editors at publishing houses that rejected proposals for this book. The challenges of book publishing are discussed in Chapter 7.

"*Sex, Drugs, Einstein, and Elves* is quite a book and fully a close encounter with your amazing mind. The book made me wish I could spend a few hours with you just picking your brain. Yes, the book is a *tour de force*."

—Science Editor, Harvard University Press

"This is truly one of the most unique ideas I've seen in a long time…"

—Adams Media

"I enjoyed reading your new proposal and am almost ready to move to Shrub Oak."

Princeton University Press

"We have never met, but reading your book sample…makes me feel that I did. Your descriptions of Shrub Oak, which I must admit I had never really seen as a nirvana, now makes me wonder whether choosing to live in Cortlandt Manor was wise.…I did enjoy your submission. I admire your chutzpah in using the Princeton University Press quote in your submission."

—Columbia University Press

"We appreciate your surrealistic take on culture, and your no holds barred approach to a memoir."

—Chronicle Books

"I'm sure there will be others who are fascinated by the associations and connections that you make."

Simon & Schuster

"The material is certainly interesting—and in some strange way, I quite enjoyed reading it."

—Berkley Books

"I share the enthusiasm of the other readers for your rather quirky musings. Very enjoyable reading."
—Stanford University Press

"I found it to be quite a ride…"
—Penguin Books

"Lots of fun…Ideas fizzing all over the place…"
—Cornell University Press

"I was honored to receive a copy of your proposal for *Sex, Drugs, Einstein, and Elves*…I have followed your publishing activities over the past 15 years…"
—MIT Press

"You have a remarkable publishing track record, and I have no doubt that you will draw publicity for this book too.
—Harcourt, Inc.

"I bet this book will get snapped up by someone."
—BenBella Books

"It looks intriguing…"
—Knopf

"I love your work…"
—Beacon Press

"…It is a very interesting project…"
—Yale University Press

"I really enjoyed reading it, by the way."
—Oxford University Press

"Reading *Sex, Drugs, Einstein, and Elves* induced a marked, noticeable psychoactive response. The author has created a text that could well result in Cliff Pickover's writing being classified as a controlled substance."
—Jamie Forbes, New York City media consultant

"It cannot be that our life is a mere bubble,
cast up by eternity to float a moment on its waves
and then sink into nothingness.
There is a realm
where the rainbow never fades,
where the stars will be spread out before us
like islands that slumber in the ocean,
and where the beautiful beings,
which now pass before us like shadows,
will stay in our presence forever."
—George D. Prentice, *Man's Higher Destiny*, 1860

"The forces that affect our lives,
the influences that mold and shape us,
are often like whispers in a distant room,
teasingly indistinct,
apprehended only with difficulty."
—Charles Dickens

"Every blade of grass has its angel
that bends over it and whispers, 'grow, grow.'"
—Talmudic commentary, *Midrash Rabbah, Bereishis* 10: 6

Schrödinger's wave equation
describes ultimate reality
in terms of wave functions and probabilities:

$$\left[-\frac{\hbar^2}{2m}\nabla^2 + V(\vec{r})\right]\psi(\vec{r},t) = i\hbar\frac{\partial\psi}{\partial t}(\vec{r},t)$$

This book is dedicated to all those
who find the equation beautiful to ponder.[*]

This book is also dedicated to
The DMT machine elves.[†]
The elves are there,
Behind the wall,
Weaving the fabric of reality.

---

[*]Physicist Freeman Dyson lauds this formula that represents a stage in humanity's grasp of reality: "Sometimes the understanding of a whole field of science is suddenly advanced by the discovery of a single basic equation. Thus it happened that the Schrödinger equation in 1926 and the Dirac equation in 1927 brought a miraculous order into the previously mysterious processes of atomic physics. Bewildering complexities of chemistry and physics were reduced to two lines of algebraic symbols."[1]

[†]These elflike beings are frequently seen by people under the influence of the psychoactive compound DMT (dimethyltryptamine).

Contents

"Readers of *Sex, Drugs, Einstein, and Elves* can fire up the old neurons by absorbing this book. This creative thinking can then be incorporated into the reader's life. As readers traverse the book, they will experience biochemical, psychophysiologic, and neurologic changes. Readers will then invoke these changes as a practical means of creative problem solving."
—Jamie Forbes, New York City media consultant

On Fugu Sushi and Transdimensional Reality Worms

In which we encounter zombie recipes, the dream fish of Norfolk Island, hallucinogenic worm intestines, fugu sushi, Homer Simpson, the "larval beings" of William Burroughs, H. P. Lovecraft, parallel universes, ayahuasca, *Attack of the Fifty Foot Woman*, the Leather Man, the Malalis Indians, grilled fugu sperm, mushroom-loving reindeer, iboga, the instinct to alter consciousness, Christian Ratsch, Wernher von Braun, Kip Thorne, Carl Sagan, *The Lobotomy Club*, the overlap of science fiction and science, and personal anecdotes of my childhood and home town of Shrub Oak, New York.

The real job of dream fish is to transport us to new seas, while deepening the waters and lengthening horizons.[1]

In which we encounter the Sapir-Whorf Hypothesis, the language of the Tariana, the karass and granfaloon, shibumi, Joumana Medlej, Alexandra Aikhenvald, Kawesqar, secret Hebrew names for God, golems, π: the movie, language as a virus; rabbis Hanania ben Teradion, Judah Low ben Bezalel and Schneur Zalman of Lyadi; beings with five billion mouths, Ös, high-IQ genius Chris Langan, "The Great Eskimo Vocabulary Hoax," time travel, Douglas Adams, Temple Grandin, Freeman Dyson, Jean-Francois Champollion, the Rosetta Stone, Arthur C. Clarke, Kurt Vonnegut, the Tetragrammaton, 666, *Lord of the Rings*, Kabbalah, free will, William S. Burroughs, Terence McKenna, aboriginal Pintupi, and the Guugu Yimithirr.

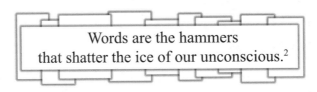

Words are the hammers
that shatter the ice of our unconscious.[2]

In which we encounter the terraqueous chrysoprase, Bertrand Russell's twenty favorite words, Piers Anthony's ten favorite words, logophiliacs, Oulipoian states of mind, constrained writing, Pynchonomancy, Annie Sprinkle, Robert Sawyer, divination, the mystery of italics, Brion Gysin's Dream Machine, music, painting, plot aesthetics, Dia Center for the Arts, lipograms, *The Anagrammed Bible*, Michael Shermer, "A Glass Centipede," computer poetry, creativity machines, Rachter, Ray Kurzweil, Yggdrasill, alien beauty, alien pornography, Reverend Jerry Falwell, adipocere, Monongahela, Harlan Ellison, Antonin Artaud, Aldus Manutius, the world's largest vocabulary, and amphigory.

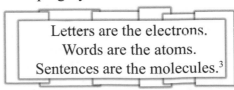

Letters are the electrons.
Words are the atoms.
Sentences are the molecules.[3]

In which we encounter Humphry Osmond, Aldous and Laura Huxley, DMT machine elves, higher-dimensional dream palaces, the Glass Chrysanthemum, Terence and Dennis McKenna, Rick Strassman, Debra Fadool, reality-enhanced mice, Whitley Strieber, pineal glands, Biblical prophets, ayahuasca, entheogens, Benny Shanon, harmaline alkaloids, Daniel Pinchbeck, robots, larval beings, alien space insects, androids, clowns, dwarves, praying-mantis entities, fairies, cities of Lego with writhing monster squids, Alexander Shulgin, Peter Meyer, Isaac Newton, William James, John Horgan, iboga, Alan Watts, Stanislav Grof, James Kent, near-death experiences, alien senses, the Indian edible-nest swiftlet, "Apparent Communication with Discarnate Entities Induced by Dimethyltryptamine (DMT)," Howard Lotsof, transcendent cable, Ken Wilbur, and temporal lobe epilepsy.

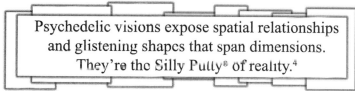

Psychedelic visions expose spatial relationships and glistening shapes that span dimensions. They're the Silly Putty® of reality.[4]

In which we encounter the illusion of intermetamorphosis, cryonics, delusional parasitosis, perception puzzles, doppelgängers, asparagus, bipolar disorder, Kay Jamison, creativity, Special K, angels, Irish Fairy and Folktales, *Invasion of the Body Snatchers*, and psychiatric disorders such as Charles Bonnet syndrome, Capgras syndrome, Fregoli syndrome, asomatognosia, prosopagnosia, reduplicative paramnesia, the syndrome of subjective doubles, Cotard's syndrome, and Ekbom's syndrome.

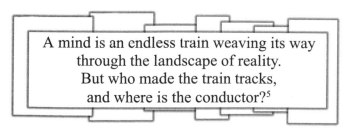

A mind is an endless train weaving its way through the landscape of reality. But who made the train tracks, and where is the conductor?[5]

In which we discuss quantum immortality, psychedelic Shakespeare, Marcel Proust, the Meseglise and Guermantes ways, Dr. Brown's Cel-Ray soda, Holiday Inn founder Kemmons Wilson, TV laugh tracks, Thomas Jefferson's decimated Bible, Warner Sallman's Head of Christ, Mel Gibson, Lucia Joyce, Carl Jung, Agent 488, General Dwight Eisenhower, Walter Benjamin and the Arcades Project, Tolkien's Lothlorien, the chanting of Sausalito mystery fish, the tenacity of the Acoemeti, recipes for attracting women, word salad and computer spam, dropping pennies from the Empire State Building, movie closing credits, near-death experiences, living in the π matrix, parallel universes, the many-worlds interpretation of quantum mechanics, Robert Heinlein, The Wishing Project, John Goddard, and the mathematical abilities of Jesus.

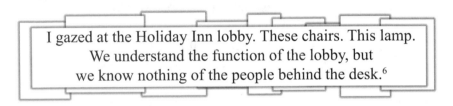

I gazed at the Holiday Inn lobby. These chairs. This lamp.
We understand the function of the lobby, but
we know nothing of the people behind the desk.[6]

In which we encounter prolific writers, literary agents, publishers, editors, bibliomaniacs, "The Whore of Mensa," Hillary Clinton, William Shatner, coffee-addict Honoré de Balzac, Dr. Seuss, cartoon guides to Proust, Isaac Asimov, publisher response times, famous rejected books, authors' advance money, lecture tours, book signings, book dedications, literary agent John Brockman, collaborations, Piers Anthony, Shirley Jackson, ghostwriters, book titles, cover art, bestsellers, writing advice, Lester Dent, Erle Stanley Gardner, Dean Koontz, Gertrude Stein, Truman Capote, Perry Mason, Willa Cather, Sigmund Freud, Ring Lardner, John Creasey, *Spider Legs*,

Pycnogonids, Jerzy Kosinski, Ellery Queen, John F. Kennedy, Agatha Christie, John Cheever, Alfred Knopf, and amphioxus.

A book is like a garden carried in the pocket.[7]

Chapter 8...197
Neoreality and the Quest for Transcendence

In which we encounter Einstein, Rumi, God, the anthropic principle, Stephen Hawking, the Bible, Proust's hyper-realities, *The Lobotomy Club*, *Sushi Never Sleeps*, fractals, brain surgery, Italian filmmaking, neorealism, stellar nucleosynthesis, the Big Bang, Paul Davies, Frank Tipler, Marcel Proust, H. P. Lovecraft, Andrei Linde, Sir Fred Hoyle, Rudy Rucker, Robert Jastrow, The Templeton Foundation, multiple universes, Paul Kammerer, synchronicity, and the shoreless sea of love.

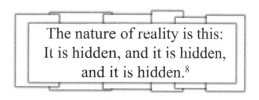

The nature of reality is this:
It is hidden, and it is hidden,
and it is hidden.[8]

Chapter 9...205
Oh God, Einstein's Brain and Eyes Are Missing

In which we encounter Einstein's brain, Maja Einstein, the Sylvian fissure, Einstein's disembodied cyes, Dr. Henry Abrams, Einstein's disembodied hair, Einstein's pipe, the legend of Mochaoi, Henri Bergson, spacetime, time dilation, Marcel Proust, Thomas De Quincey, Angelus Silesius, Félirena Naomh Ncrennachin, Philip K. Dick, Mount Coelian, the Seven Sleepers, God's time, Moslem legends, magical horse Burak, Isaac Newton, Hollywood, Marian Diamond, Stephen Hawking, the Chronology Projection Conjecture, psychoactive ketamine, the special theory of relativity, *In Search of Lost Time*, and future-life progression. Is time travel possible? If time is something learned, can we unlearn it? Can the flow of time be stopped? What if Einstein had lived another twenty years?

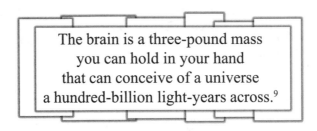

The brain is a three-pound mass
you can hold in your hand
that can conceive of a universe
a hundred-billion light-years across.[9]

In which we encounter Ts'ai Lun and the invention of paper, David Jay Brown's *Conversations on the Edge of the Apocalypse*, nanotechnology gray goo, rise of the machine civilization, evolution of the human race, ring angels, insectile aeroplankton, dragonflies, George Bush, zygotic personhood, abortion, stem cells, Alan Dershowitz, Kilgore Trout, the extinction of the Jews, DMT machine-elf research centers, Jovian moons, angelic beings, Burning Man and the Aortic Arch, supermodels, the Superbug Age, musicians and monks, The Bone Tree, The Nebulous Entity, The Plastic Chapel, Xeni Jardin, Jeff Bezos, John Brockman, Maria Spiropulu, Connie Willis, Stephen Spielberg, Arthur C. Clarke, Freeman Dyson, Neal Stephenson, Alexandra Aikhenvald, Amy Chua, Maggie Balistreri, the Lucidity Institute, Jane Roberts, Mr. Spock, Whitley Strieber, extraterrestrials, Star Wars, intelligence, evolution, hive minds, termites, the Turing test, Chworktap, Dean Koontz, extinction of the human species, Dr. Rick Strassman, Terence McKenna, doomsday machines, Freeman Dyson, and Tinkertoy minds. Can consciousness exist without a brain?

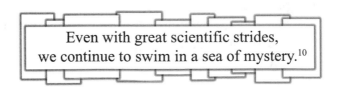

Even with great scientific strides,
we continue to swim in a sea of mystery.[10]

Chapter 11...251
Farewell

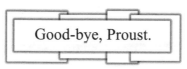

Good-bye, Proust.

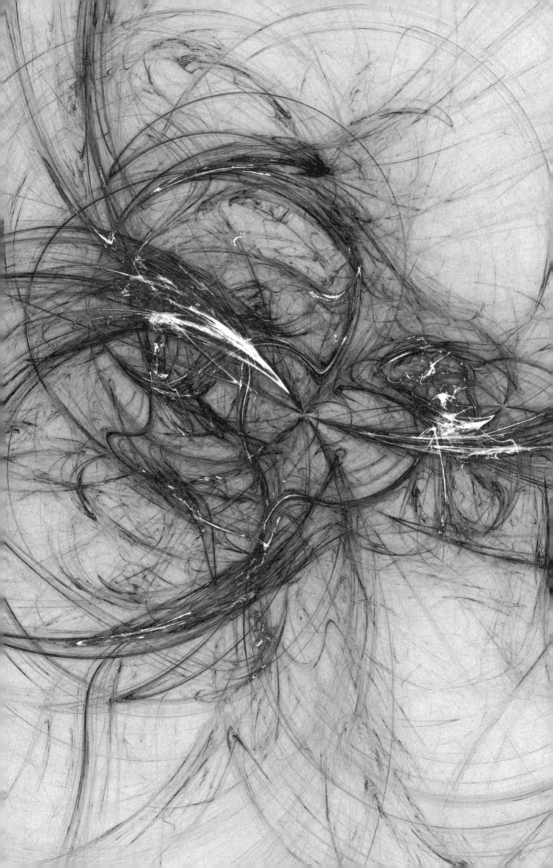

Acknowledgments

I thank Mark Nandor, Dennis Gordon, Lon Richardson, Teja Krasek, Jamie Forbes, David Jay Brown, Graham Cleverley, Pete Barnes, Steven Andersen, and members of the Clifford Pickover Think Tank for useful comments and suggestions:

http://groups.yahoo.com/group/CliffordPickover/

Robert Hendrickson's *The Literary Life* is an excellent source of anecdotes on the subject of books, authors, and publishers. Many of the Einstein quotations come from Alice Calaprice's *The Expanded Quotable Einstein*. Information on Marcel Proust comes from diverse sources such as Stephane Heuet's graphic novels on Proust's works, Roger Shattuck's *Proust's Way*, Edmund White's *Marcel Proust*, and Alain de Botton's *How Proust Can Change Your Life*. Other influential books, such as those that discuss the psychoactive compound DMT and uncommon psychiatric disorders, are listed in the Further Reading section at the end of this book.

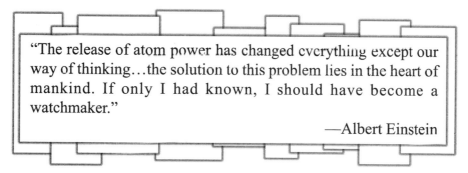

"The release of atom power has changed everything except our way of thinking...the solution to this problem lies in the heart of mankind. If only I had known, I should have become a watchmaker."

—Albert Einstein

Preface

In the summer of 2004, I became especially curious about humanity's age-old quest for transcendence through religion, architecture, art, psychedelics, language, and countless other means of expression. A passion scized me, and I resolved to explore humankind's greatest examples of creative fervor.

I explored the ruins of Pompeii in Italy and walked the narrow streets of Eze, a medieval mountain city in France. Shivers ran through me as I knelt in Saint Peter's Basilica and in front of the Black Madonna in the mountain monastery of Montserrat. I rested in the Miracle Field of Pisa and in cafes on the terraced, narrow streets of Villefranche. I stood only inches from the bones of Michelangelo and Galileo in Florence's Church of Santa Croce.

Fully energized, I wandered through the Roman Coliseum, the ruins of Imperial Rome, the psychedelic Sagrada Familia Cathedral in Barcelona, the Piazza del Duomo in Florence, and the Notre-Dame de la Garde—a beautiful basilica set high atop the hills above the Old Harbor in Marseille.

One of the major themes of this book jelled as I toured the cliff-side town of Positano on the Amalfi coast near Naples, where buildings grow like barnacles—from the tops of hills and down the sheer slope to the Gulf of Salerno. It suddenly hit me that behind every society was a hidden, elflike voice that whispered: "Build! Create! Build! Create!" Moreover, some of the intricate structures I had seen during my journey were reminiscent of the sparkling, ornate palaces revealed to people under the influence of the psychoactive compound DMT (dimethyltryptamine). It seems as if DMT frees the mind to see the blueprint—hardwired by the whispering elves—instructing us to create, create, create.

My fascination with people's DMT visions leads me to discuss DMT "realities" in great detail throughout this book. I also dwell on Marcel Proust and Albert Einstein, perhaps the ultimate expressors of creativity in literature and science. Just as termites are designed to make intricate mounds, Golden Orb Web spiders to weave tremendous webs, and bower birds to construct ornate nests decorated with colorful baubles including feathers, berries, pebbles, and shells—so our species is designed to build magnificent jeweled palaces, seek glitter, and compose symphonies.

Mediterranean travel reinforces the notion that architecture is the quintessence of creativity and practicality. In order for any building to get off the ground—literally—the architect must not only have a sense of art and of creating new spaces, he or she must consider new technologies, materials, engineering constraints, politics, zoning laws, and impact on communities. Architecture—a pinnacle of creative expression tempered by basic human needs and limitations—is emblematic of the entire range of topics covered in the book that require readers to reason and to dream.

I completed this book on my return to the United States, while walking through the rustic streets of my home town, taking notes, resting, and enjoying the country air. These are my personal musings on our quest to understand reality and to transcend our ordinary lives.

Ever since the Tower of Babel, humans have built, created, and sought beauty. Alas, the Tower makers were scattered, their language confounded, and the Tower abandoned. In this book, let's rebuild some of the lesser-known staircases of the Tower. After all, even after our expulsion from Eden and the flight from the Tower, God gives us back our creativity in Exodus 35. He gives it all to a man named Bezalel of the tribe of Judah, and Bezalel opens the gates for all humankind:

> Moses said to the children of Israel: "See, the Lord has called by name Bezalel…and has imbued him with the spirit of God, with wisdom, with insight, and with knowledge, and with talent for all manner of craftsmanship to do master weaving, to work with gold, silver, and copper, with the craft of stones for setting and with the craft of wood…And He put into his heart the ability to teach…He imbued him with wisdom of the heart, to do all sorts of work of a craftsman and a master worker and an embroiderer with blue, purple, and crimson wool, and linen and of weavers…Bezalel and Oholiab and every wise-hearted man into whom God had imbued wisdom and insight to know how to do, shall do all the work of the service of the Holy, according to all that the Lord has commanded."[1]

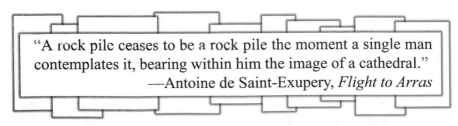

"A rock pile ceases to be a rock pile the moment a single man contemplates it, bearing within him the image of a cathedral."
—Antoine de Saint-Exupery, *Flight to Arras*

"Imagination is more important than knowledge. For while knowledge defines all we currently know and understand, imagination points to all we might yet discover and create."
—Albert Einstein, "On Science"

"Somewhere in that great ocean of truth, the answers to questions about life in the universe are hidden...beyond these questions are others that we cannot even ask, questions about the universe as it may be perceived in the future by minds whose thoughts and feelings are as inaccessible to us as our thoughts and feelings are to earthworms."
—Freeman Dyson, "Science & Religion: No Ends in Sight,"
New York Review of Books

Introduction

"Where did it begin? Was there nothing and then something? Even the Celestials, the noblest of beings to enter the human consciousness, did not know what was in the dark before the light was turned on."
—Peter Lord-Wolff, *The Silence in Heaven*

Dreaming at the Edges of Reality

I don't believe in astrology. I don't believe that aliens are descending to create crop circles or mutilate cows. I don't believe in a personal God who smites the wicked, or in TV psychics who can read minds or talk to dead Aunt Sally. I don't believe that Moses parted the Red Sea like Charlton Heston in *The Ten Commandments*. I don't believe that the Earth is hollow, and that inside lives an advanced civilization known as the Agartha.

I'm so skeptical that I've written numerous articles for American's foremost skeptic magazine, *Skeptical Inquirer*. On the other hand, unlike some skeptics I know, I am deeply fascinated by religion, near-death experiences, psychoactive compounds, the spiritual, and even the Bible.

Consider the Bible as just one example of my obsessions. To me, the Bible is an alternate-reality device. It gives its readers a glimpse of other ways of thinking and of other worlds. It is also among the most mysterious books ever written. We don't know the ratio of myth to history. We don't know all the authors. We are not always sure of the intended message. We only know that the Bible reflects and changes humankind's deepest feelings. The Bible chips away at the edges of our unconscious. At a minimum, the Bible is a fascinating model of human understanding—of how we reach across cultures to connect with one another and learn about what we hold as sacred. Perhaps with this kind of interface, one might even look for realms beyond our ordinary reality.[1]

This book is all about reality. Warped reality. Shattered reality. Funny reality. I hope that the smorgasbord of topics in this book help bend reality for you, or at least help you see the world in a fresh way. Read on. I want to make you laugh and stretch your mind.

This book meanders from one far-flung topic to another to test your curiosity and powers of lateral thinking. Robert Pirsig wrote in *Zen and the Art of Motorcycle Maintenance*, "It's the sides of the mountain which sustain life, not the top. Here's where things grow." This also applies to the joy that writers experience when letting their minds drift and when wondering about humanity's place in the universe. To this end, *Sex, Drugs, Einstein, and Elves* is a collection of personal essays on topics I contemplated after my Mediterranean journey, while walking through Shrub Oak, New York, my home town. Subjects include fugu sushi, zombies, French writer Marcel Proust, parallel universes, DMT-induced realities, hallucinogenic worms, Holiday Inn founder Kemmons Wilson, Thomas Jefferson's decimated Bible, religious states, uncommon psychiatric disorders, drug-induced visions, Lucia Joyce, Warner Sallman's *Head of Christ,* Albert Einstein, shamanist Terence McKenna, Burning Man, New York literary agents, quantum theory, and the humming toadfish, whose incessant underwater droning at a perfect A-flat was a mystery for years.

Shrub Oak is very tame, but I find that wandering through tree-lined streets—past ancient churches and the rustic grounds of old libraries—helps me think and ponder the weird. There are no elves in Shrub Oak, as far as I can tell. The "elves" in this book's title refer to entities that people see while taking the psychoactive chemical DMT or while in other altered states.

I develop most ideas for my books while walking outdoors or in dreams at night, which is pretty common for authors and scientists. For example, Indian mathematician Srinivasa Ramanujan (1887–1920) loved dreams for gaining mathematical insight. He was an ardent follower of several Hindu deities. After receiving nightly visions in the form of droplets from these gods, Ramanujan saw scrolls containing very complicated mathematics. When he woke from his dreams, he wrote down only a fraction of what the gods showed him. Robert Kanigel, in *The Man Who Knew Infinity*, suggests that Ramanujan's melding of spirituality and mathematics signified a "peculiar flexibility of mind, a special receptivity to loose conceptual linkages and tenuous associations...." Indeed, Ramanujan's openness to mystical visions suggested "a mind endowed with slippery, flexible, and elastic notions of cause and effect that left him receptive to what those equipped with purely logical gifts could not see."[2]

As I note in my book *A Passion for Mathematics,* Ramanujan was a prophet who plucked mathematical ideas from another world. They came to him in a flash. He could read the codes in the mathematical matrix in the same way that Neo, the lead character in the movie *The Matrix,* could access mathematical symbols cascading about him that formed the infrastructure of his perceived reality. I don't know if God is a cryptographer, but codes are all around us waiting to be deciphered. Some may take a thousand years to understand. Some may always be shrouded in mystery.

Throughout history, creative people have been open to dreams as a source of inspiration. Paul McCartney said that the melody for the famous Beatles' song "Yesterday," one of the most popular songs ever written, came to him in a dream. Apparently the tune seemed so beautiful and haunting that for a while he was not certain it was original. Danish physicist Niels Bohr conceived the model of an atom from a dream. Elias Howe received in a dream the image of a needle design required for a lock-stitch sewing machine. René Descartes was able to advance his geometrical methods after flashes of insight that came in dreams. The dreams of Dmitry Mendeleyev, Friedrich August Kekulé, and Otto Loewi inspired scientific breakthroughs. It is not an exaggeration to suggest that many scientific and mathematical advances arose from the stuff of dreams.

Decades after Ramanujan, many of the modern world's greatest minds have been inspired by marijuana and psychedelics. As just one example, Ralph Abraham, a pioneer in chaos theory and Professor of Mathematics at the University of California, explained how psychedelic insights influence mathematical theories:

> In the 1960s a lot of people on the frontiers of math experimented with psychedelic substances. There was a brief and extremely creative kiss between the community of hippies and top mathematicians. I know this because I was a purveyor of psychedelics to the mathematical community.
>
> To be creative in mathematics, you have to start from a point of total oblivion. Basically, math is revealed in a totally unconscious process in which one is completely ignorant of the social climate. And mathematical advance has always been the motor behind the advancement of consciousness.[3]

Playing at the Edges of Reality

My colleagues sometimes wonder why I'm so curious about the fringes of science and about smart people who play at the borderlands of science. I believe that "fringe" research is crucial—not just for its educational value but because significant discoveries can come from such study. At first glance, some topics in science or sociology may appear to be curiosities, with little practical application or purpose. However, I have found these experiments useful and educational, as have the many students, educators, artists, and scientists who have written to me.

Play is important everywhere in science. Although this book is not one of my more traditional "science books," I should emphasize that many important breakthroughs in science have been discovered accidentally. In particular, science is filled with hundreds of great discoveries that have emerged through chance happenings and serendipity, for example: Velcro, Teflon, X-rays, penicillin, nylon, safety glass, sugar substitutes, dynamite, and polyethylene plastics.

In this book, we go beyond traditional technology and science and delve into parallel realities, religion, cultural curiosities, and strange beings. Whatever we believe about such weird topics, we must admit that, throughout history, humans have often experienced feelings and ideas that transcended their ordinary lives. These experiences are not always regarded as divine, useful, or "true." Psychiatrists may relegate them to heightened activity of the brain's temporal lobes, though this doesn't necessarily make the insights and visions invalid. Psychologist William James has argued that religious states are not less profound simply because they can be induced by mental anomalies:

> Even more perhaps than other kinds of genius, religious leaders have been subject to abnormal psychical visitations. Invariably they have been creatures of exalted emotional sensitivity liable to obsessions and fixed ideas; and frequently they have fallen into trances, heard voices, seen visions, and presented all sorts of peculiarities which are ordinarily classed as pathological.[4]

James goes on to say that these pathological features of "genius" have helped to give them their religious authority, influence, and wisdom. To James, "abnormal" brain states provide valid ways for religious figures to gain insight:

To plead the organic causation of a religious state of mind in refutation of its claim to possess superior spiritual value, is quite illogical and arbitrary because none of our thoughts and feelings, not even our scientific doctrines, not even our *dis*-beliefs, could retain any value as revelations of the truth, for every one of them without exception flows from the state of the possessor's body at the time...Saint Paul certainly once had an epileptoid, if not an epileptic, seizure, but there is not a single one of our states of mind, high or low, healthy or morbid, that has not some organic processes as its condition.[5]

Marcel Proust and Hyperreality

Book critics, beware! I ruminate and wander freely through a vast carnival of topics, seizing every opportunity to digress and explore mental tributaries. This reminds me of the quote from *Don Quixote*, which has become a guiding principle in my life: "He calmly rode on, leaving it to his horse's discretion to go which way it pleased, firmly believing that in this consisted the very essence of adventure." I'm no Don Quixote, but I agree with French novelist Marcel Proust (1871–1922) when he asserts that great works of art, and significant books, have little to do with their subject matter, but more to do with the treatment of that matter. He thought *everything* was a fertile subject for books. He peered into all aspects of mind, the fringes or reality, the beauty of nature, and the quirks of pop culture of his time. He used soap advertisements and train schedules as seeds for profound ideas. I hope you agree with Proust and value my own eclectic assortment of topics.

When I wake up at night with ideas, I jot the ideas on a pad so that I can recall them in the morning. Proust also woke with his head flooded with memories and ideas. He sometimes awoke unaware of where he was or even who he was. He seemed to transcend space and time as he remembered events of his youth or conjured poetic visions of his great-aunt's house in Combray or his trips to Paris, Doncières, and Venice. As you will see through anecdotes scattered throughout this book, his realities are often not ordinary realities to which you and I have ready access.

Proust is my hero, and even though he did not have formal scientific education as I have, his life had intimate resonances with science and psychology. The *New York Times* recently proclaimed: "Like Einstein, Marcel Proust was, in his own way, a theorist of time and space."[6] Proust wrote, "An hour is not merely an hour—it is a vase full of scents and sounds and projects and climates."[7] We'll discuss Proust in the following chapters and see how the mere taste of a madeleine cookie triggered a cascade of childhood memories that consumed his adult life. I find it amazing that his epic work *Remembrance of Things Past* (also translated as *In Search of Lost Time*) continues to grow in popularity, has never been out of print, and penetrates today's popular culture. He was even mentioned in an episode of the hit TV drama *The Sopranos*, where mobster Tony Soprano's psychiatrist, Dr. Melfi, mentions Proust as Tony searches for the meaning of his life:

> *TONY:* All this from a slice of gabagool [capicola ham]?
> *DR. MELFI:* Kind of like Proust's madeleines.
> *TONY:* What? Who?
> *DR. MELFI:* Marcel Proust. Wrote a seven-volume classic, *Remembrance of Things Past*. He took a bite of a madeleine—a kind of tea cookie he used to have when he was a child—and that one bite unleashed a tide of memories of his childhood and ultimately, his entire life.
> *TONY:* This sounds very gay. I hope you're not saying that.[8]

In other words, capicola, the Italian spiced ham, emerges as Tony Soprano's equivalent of Proust's madeleine cookie for triggering memories and insights. For me, the trigger is a walk along Main Street in Shrub Oak. Thus, as I've said, many of the topics in this book percolate from memories jotted down in a notebook, while I walk tree-lined streets.

Reality Carnival

The smorgasbord of fun and quirky facts, questions, anecdotes, and puzzles in this book are metaphors for society and science. You can pick and choose from the various delicacies as you explore the platter set before you. Some of the topics are personal and deal with my interactions with my agents and publishers. Others revolve around psychedelic experiences that suggest physical reality is more pliable than you may currently believe. Several chapters of this book discuss powerful, illegal drugs, and I do not advocate or suggest that readers break any laws or try these drugs. Many topics are arranged to enhance the sense of adventure and surprise: My brain is a runaway train, and these anecdotes and topics are the chunks of my cerebrum strewn on the tracks.

Occasionally, some of the topics or puzzles in this book will seem simple or frivolous, including dropping pennies from the Empire State Building or wondering if Jesus could calculate 43×31. However, these are questions that fans have often posed to me, and I love some of these "quirkies" the best. I agree with the Austrian physicist Paul Ehrenfest who said: "Ask questions. Don't be afraid to appear stupid. The stupid questions are usually the best and hardest to answer. They force the speaker to think about the basic problem."[9]

I agree with Ehrenfest and particularly enjoy questions from children. But I sometimes wonder if scientists will be able to answer humanity's deepest questions. Perhaps our brains, which evolved to make us run from lions on the African savanna, are not constructed to penetrate the fabric of the universe. Imagine an alien with an IQ a hundred times our own. What profound concepts might be available to this creature in areas of awareness to which we are now totally closed? The average poodle cannot understand Fourier transforms, the Sapir-Whorf hypothesis, or gravitational wave theory. Human forebrains are a few ounces bigger than a poodle's, and we can ask many more questions than a poodle. Are there facets of the universe we can never know? Are there questions we can't even ask? Michael Murphy discusses a related idea in *The Future of the Body*:

> To a frog with its simple eye, the world is a dim array of grays and blacks. Are we like frogs in our limited sensorium, apprehending just part of the universe we inhabit? Are we as a species now awakening to the reality of multidimensional worlds in which matter undergoes subtle reorganizations in some sort of hyperspace?[10]

Although I mention Albert Einstein throughout the book, you won't encounter a lot of deep theory relating to Einstein. Instead, I invoke his name metaphorically to call up the mystery and the joy that many researchers feel when tinkering at the edges of science. Certainly, Einstein played at the edges, contemplating such concepts as time travel, warped spacetime, and higher dimensions. In a 1929 interview, Einstein sets the tone for *Sex, Drugs, Einstein, and Elves*:

> We are in the position of a little child entering a huge library whose walls are covered to the ceiling with books in many different

tongues…The child does not understand the languages in which they are written. He notes a definite plan in the arrangement of books, a mysterious order which he does not comprehend, but only dimly suspects.[11]

A quote by John Steinbeck from *The Sea of Cortez* gives me permission to play:

The design of a book is the pattern of reality controlled and shaped by the mind of the writer. This is completely understood about poetry or fiction, but is too seldom realized about a book of facts.[12]

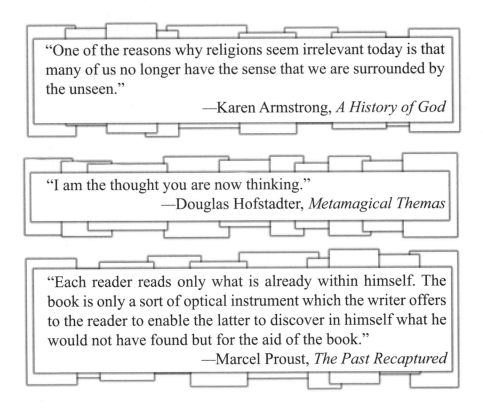

"One of the reasons why religions seem irrelevant today is that many of us no longer have the sense that we are surrounded by the unseen."
　　　　　　　　　　—Karen Armstrong, *A History of God*

"I am the thought you are now thinking."
　　　　　　　　　—Douglas Hofstadter, *Metamagical Themas*

"Each reader reads only what is already within himself. The book is only a sort of optical instrument which the writer offers to the reader to enable the latter to discover in himself what he would not have found but for the aid of the book."
　　　　　　　　　　—Marcel Proust, *The Past Recaptured*

On Fugu Sushi and Transdimensional Reality Worms

In which we encounter zombie recipes, the dream fish of Norfolk Island, hallucinogenic worm intestines, fugu sushi, Homer Simpson, the "larval beings" of William Burroughs, H. P. Lovecraft, parallel universes, ayahuasca, Attack of the Fifty Foot Woman, *the Leather Man, the Malalis Indians, grilled fugu sperm, mushroom-loving reindeer, iboga, the instinct to alter consciousness, Christian Ratsch, Wernher von Braun, Kip Thorne, Carl Sagan,* The Lobotomy Club *the overlap of science fiction and science, and personal anecdotes of my childhood and home town of Shrub Oak, New York.*

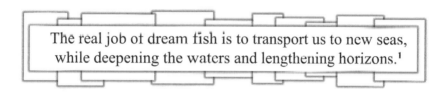

The real job of dream fish is to transport us to new seas, while deepening the waters and lengthening horizons.[1]

My Life in a Nutshell

Shrub Oak is only an hour north of New York City, but it might as well be in an entirely different universe. The town has little pollution, no skyscrapers, few honking horns. The pride and joy of Shrub Oak is its bucolic Main Street—with the elegant John C. Hart Memorial Library, cottage-style homes, and landscaping with meandering pathways, benches, water fountains, and narrow ivy-covered passageways between quaint stores. I've always been happy here. Much of this book results from notes I have made while walking down Main Street.

I first became interested in mystical experiences and warped realities while eating sushi and contemplating psychedelic worm intestines. Let's discuss the psychedelic worms in a moment, because first I would like to tell you about my childhood and where I live today.

I was born in Red Bank, New Jersey, in 1957. At that time, my father worked at the Bendix plant in Eatontown on Highway 35. A year after I was born, he got a job at Fort Monmouth where he became the chief of a department of the Communication Electronic Communication Command (CECOM). His team was responsible for designing electronic equipment as well as placing and monitoring military-contractor programs. My mother taught first grade in local schools. My one sibling, Larry, is two years younger than I. He still lives in New Jersey and is a brilliant gastroenterologist. He spends his days placing long, snaking tubes into people's rectums.

My childhood was happy, conventional, and middle class. We lived in a small house in Shorecrest, New Jersey, just a few blocks away from Highway 35. I remember our split-level home at 39 Richard Terrace and the nearby Bodman Park where I used to play. Both parents helped me with my homework as I grew up—my dad worked with me on mathematics, and my mom helped me with presentations and posters. My father also continually drew mazes for me to solve with pencil and paper.

I fondly recall watching the old black-and-white *Outer Limits* and *Chiller Theater* shows on Saturday-night TV, which featured such notables as *Attack of the Fifty Foot Woman* and *Attack of the Crab Monsters*. (For film buffs, *Crab Monsters* was made in 1957 for $70,000, and grossed approximately $1 million, making it Roger Corman's most profitable picture of the period.) Oh God! I can still remember the scenes in which scientists discover a pair of giant crabs mutated by atomic tests on a remote island.

After second grade at Fairview Elementary School, we moved to a bigger home in Ocean Township, New Jersey, where I later completed high school with a class rank of 2. From there, I went to Franklin & Marshall College, graduating with a class rank of 1 in three years. Ever onward, I went to Yale University, where, in 1982, I received my Ph.D. from the Department of Molecular Biophysics and Biochemistry. The impressive "Yale" name was useful to gain people's respect later in life, but the actual education I got at the smaller Franklin & Marshall College was superior. Professors at the biggest-name schools are sometimes more interested in their research than in teaching.

After Yale, I landed a job at the IBM Watson Research Center in Yorktown Heights, New York. A few years later, I got married and had a son. My wife, an Iranian Baha'i, came to America with her family about the time Ayatollah Khomeini rose to power. As an aside, because more than 50 percent of Jews will marry non-Jews, as I did, American Jews will essentially vanish sometime in this century, except for fecund enclaves of the religious Hasidim. This extrapolation is based on the tendency for children of mixed marriages to be raised as non-Jews.

Judaism is also coming to an end in Europe. In 1946, there were four million Jews. Today, there are two and a half million. Ireland will be the first to lose its Jewish community altogether. Rather than become depressed about the loss of the great Jewish culture, I hope that other religions, like the Baha'i faith, will serve to elevate humanity in the coming centuries.

While on the topic of religion, I should mention that I've always wanted to write books on God and religion, and did so with such titles as *The Paradox of God and the Science of Omniscience* and *The Loom of God*. However, my science books have generally sold better than my God books. For example, my first book, *Computers, Pattern, Chaos and Beauty*, published in 1990, was one of my biggest sellers, despite its numerous equations. My 30th book was published in 2004.

And that's my life story in a nutshell.

I've been fascinated by science since childhood. While growing up in New Jersey, my bedroom featured plastic anatomical models of the heart, brain, and eye; posters of the human circulatory system; trilobite fossils, science-fiction books, and Ugly Stickers displaying wild-eyed, grinning creatures with names like "Bob, "Sandy," and "Iris." I recall trekking to a small grocery store called Emmonds on Highway 35 to get a pack of stickers, which included a bubble gum stick. I still have the stickers today.

My childhood interest in science arose from my desire to learn how the world works and from my passion for science fiction. As a teenager, one of my favorite science-fiction tales was Henry Hasse's "He Who Shrank," originally published in 1936, which describes the exploration of subatomic universes filled with machine civilizations. Many scientists and science popularizers got kick-started in life by reading science fiction.

Today my taste for home decorations has morphed from bedroom anatomy posters to wooden African masks that currently surround a spiral staircase near the dining room. These masks are modern works, some from the early part of the 20th century, some contemporary. Portraiture is one of the great universal traditions in art: the face as expressed in wood is eminently versatile, and a wonderfully direct way to connect with the viewer. My masks spill out into the dining room, complete the frame of the hallway, and threaten to overrun the living room area. At this rate of growth, I estimate that by 2015, no empty square inch of wall will remain in the entire house.

Shrub Oak, New York

Last night, I enjoyed "Bakudan Maki roll"—exotic sushi comprising crisp sweet shrimp, baby arugula, caviar, curried coconut, and monkfish liver. Sushi is a delicacy I relish above all others, and I eat an eclectic assortment every few days in one of the several local Japanese restaurants in Yorktown, New York. I do my best thinking while eating sushi or walking along Main Street after a meal. Indeed, my novel *Liquid Earth* takes place entirely on the shady street. This is my sanctuary, my suburban cathedral.

Leaf-strewn Main Street winds through Yorktown's country hamlet of Shrub Oak—two miles of road that passes ancient cemeteries, the Victorian-style Hart library, and columned homes dating back to the early 1900s. My town was first settled in the late 17th century and was predominantly a farming community into the 20th century. The most colorful person to have his name associated with Shrub Oak was "town tramp" Jules Bourglay, better known as "The Leather Man." He made regular stops in Shrub Oak in the late 1880s while traveling on foot between the Hudson and Connecticut rivers. According to our local historian, "The legendary Leather Man was never known to speak, and led a solitary existence, living in caves."[2] The Leather Man was a tenuous and mysterious bridge from my familiar realm of Shrub Oak to the world beyond.

The Leather Man has long since turned to dust. In the 1930s, Shrub Oak became a popular summer destination for families in New York City and lower Westchester. I landed here in the early 1980s, after graduating from Yale. Sushi didn't come to our town until a few years ago.

Every now and then I have difficulty drifting to sleep, wondering what became of the Leather Man. But mostly I sleep pretty well.

Fugu Sushi

Marcel Proust, the French writer I will frequently mention, used to look at train timetables to help him fall asleep. The exotic names of rural villages on the train schedule stimulated his imagination to conjure entire villages with colorful inhabitants. The names of northern railway stations in the timetable allowed him to imagine himself "stepping from the train on an autumn evening, when the trees are already bare and smelling strongly in the keen air [that is] full of names he has not heard since childhood."[3] For Proust, a train schedule could be more important than "fine volumes of philosophy."

My personal "Proustian" stimulus of the imagination is the sushi menu. The mere names of the sushi—ikura, sayori, tamago—along with their artful displays, provide my mind with a splash of ideas, ranging from biology to art to Asian travel. I may have taken this sushi obsession a bit too far when I named all the smart, punk women in my novel *The Lobotomy Club* after items on a sushi menu. Sayori, the main character, has an artificial arm. But I digress, as usual.

My fascination with sushi is amplified by the expenses sometimes involved. A rich friend of mine regularly visits Masa Takayama's sushi restaurant, hidden behind an unmarked door on the fourth floor of the Time Warner Center in New York City. Prices can be as high as $500 per person, plus a $100 kill-fee per person if the reservation is canceled, making this among the most expensive sushi restaurants in America! Masa's live shrimps and crabs come from Japan. The yellowtail and spicy baked oysters are served with black truffles. Patrons eat off of a $60,000 hinoki-wood sushi bar. The walls of the restaurant are covered with blocks of mottled Japanese Oya stone personally selected by Masa Takayama. According to *Los Angeles Times* food critic S. Irene Virbila, "Perfection is almost impossible in a restaurant, but Takayama somehow achieves it."

Although I love sushi, I've never tried the famous and dangerous fugu fish, preferring the safer tekka maki and yellowtail with scallions. My dream *is* to someday try fugu, and I often wonder about the accelerating appeal of this rare delicacy. Why is this fish all the rage? The raw flesh of this deadly blowfish is not supposed to be much tastier than the standard fare. My guess is that most fugu aficionados want to eat the fish because they know that it will kill them if incorrectly prepared. This is sheer adventure and something sure to impress your date, as he or she watches you place the flesh in your mouth with exponentially increasing dread or anticipation!

The best fugu in all Japan is reputed to be in the port city of Shimonoseki and comes in a variety of "styles":

- ◆ Fugu-sashi (delicate strips of raw flesh the thickness of paper)
- ◆ Fugu-no-karaage (fugu deep fried)
- ◆ Fugu-chiri (vegetables and fugu in soup)
- ◆ Fugu hire-zake (grilled fin in hot sake drink)
- ◆ Yubiki or Teppi (fugu skin with ponzu sauce)
- ◆ Whirako yaki (grilled fugu sperm)

Imagine requesting this last delicacy. "Please, Sir, I'd like to have a take-out order of whirako yaki. You see, my friends love the stuff but they're too shy to actually order it, given what it is."

The Japanese eat a dozen kinds of blowfish, but the Torafugu (tiger blowfish) is the most delicious. Some nervous Japanese approach the fish while uttering a Zenlike mantra: "I want to eat fugu, but I don't want to die." Happily for the Japanese, only licensed cooks are allowed to prepare fugu; nevertheless, dozens of deaths occur every year due to exuberant fugu consumption. Only a few deaths have been reported in the United States.

Japanese chefs tell me that fugu, when alive, is the only fish that can close its eyes. I don't know if that's true, but when you kill the fish, it does indeed appear to squint, while make a shrieking, crying sound. I once wrote a science-fiction story in which various wooden troughs of flailing fugu are laid before a sushi chef who is himself a fan of the delicacy. As he gazes at nearly fifty fish lots, he says, "There's nothing more macho than being able to eat your dinner after hearing the crying of lot 49."

Zombies

The fugu puffer fish is deadly because its sex organs and liver contain a neurotoxin called tetrodotoxin (anhydrotetrodotoxin 4-epitetrodotoxin, tetrodonic acid). Much smaller amounts of the poison can be found in the fish's flesh, and this can cause a slight tingling and feeling of euphoria when eaten. Maybe that's a reason people enjoy the fish. Tetrodotoxin has also been isolated from starfish, the California newt, frogs of the genus Atelopus, the blue-ringed octopus, angelfish, and xanthid crabs.

Anthropologist Wade Davis has hypothesized that tetrodotoxin is one of the active ingredients in voodoo "zombie powder." Haitian witch doctors may have used tetrodotoxin to poison victims in order to give them the appearance of being dead. Their paralyzed bodies would have heartbeats that are difficult to detect. After the tetrodotoxin wears off, the recipient revives and seems to awaken from the dead.

In the late 1980s and early 1990s, several scientific papers and books were published with such delicious titles as "Evidence for the Presence of Tetrodotoxin in a Powder used in Haiti for Zombification" and *Passage of Darkness: The Ethnobiology of the Haitian Zombie*. Zombification by fugu was even the topic of a second-season episode of the *X-Files* titled "Fresh Bones" in which FBI agents investigate curious deaths at a U.S. resettlement camp for Haitian refugees. My favorite episode of *The Simpsons* was the one in which Homer eats poorly prepared fugu.

> *HOMER*: Poison? What should I do, what should I do? Tell me, quick.
>
> *SUSHI CHEF*: Oh, no need to panic. There's a map to the hospital on the back of the menu.[4]

Insect Eating and Transcendence

That's enough talk about fugu. Right now, I'm carrying a pad of paper—to be more precise, a Staedtler® Engineering Computation Pad, 5 × 5 Grid. As I walk down Shuub Oak's Main Street, I've jotted several additional topics I'd like to discuss with you: (1) Hallucinogenic Worms, (2) Dream Fish, and (3) H. P. Lovecraft. Let's start with the hallucinogenic worms. My longtime fascination with topics at the edges of science started with Everard B. Britton's provocative paper, "A Pointer to a New Hallucinogen of Insect Origin."[5] This short paper sparked my curiosity about such far-flung topics as psychoactive drugs, parallel universes, and chemicals in the brain that induce visions of alien "elves" (Chapter 4) who seem to weave the fabric of reality.

Everard Britton is one of Australia's most distinguished entomologists. For more than twenty years, he was a curator at the Natural History Museum in London. Now he's retired but still retains his passion for beetles such as chafers and the Melolonthinae.

Back in 1984, Britton called my attention to the existence of a new hallucinogen, unusual because its source is an insect. The extinct Malalis natives of eastern Brazil once regarded the *bicho de tacuara* (bamboo worm) as a special culinary treat that tasted like sweet cream. The small creatures are found in bamboo stems.

Before we discuss the hallucinogenic effect of worm eating, realize that insect eating (entomophagy) is quite common in countries like Thailand and Mexico, where hundreds of species are consumed. Aristotle adored the flavor of cicadas. Outside of Europe and North America, *most* people eat some insects. In parts of Africa, more than 60 percent of dietary protein comes from insects. Grubs and caterpillars contain healthy unsaturated fats.

A friend once ate at a banquet hosted by the New York Entomological Society where the menu included chocolate cricket torte, mealworm ganoush, sautéed Thai water bugs, and waxworm fritters with plum sauce. He mentioned a movie theater in Colombia where roasted ants were eaten like popcorn. As delicious as these treats were, his favorites were honeypot ants, with their transparent abdomens distended with peach nectar.

Today, Thais love fried grasshoppers. The Japanese eat the rice grasshopper known as hashi. Africans enjoy locusts and certain caterpillars. Many people in Central and South America consume stink beetles.

Several companies ship edible insects to patrons in America. For example, Grubco, Inc. from Fairfield, Ohio, is one of the nation's leading suppliers of edible insects.[6] They tell me that human consumption is rising. Every week, company president Dale Cochran sells over 10,000 crickets, mealworms, and waxworms to people who eat insects. Why should we consider this weird? Many of us enjoy shrimp, lobsters, and other "insects of the sea" that belong to the same biological phylum, Arthropoda, that includes land insects.

Yes, many people eat insects, but who would have guessed that insects could be psychedelic portals to other universes? In the 1800s, the Portuguese who lived among the Malalis Indians also enjoyed entomophagy and "melted" the bamboo worms over fires, so that the worms formed an oily mass. During the cooking, the Portuguese had to be careful! The Malalis warned that the worm's head was a dangerous poison. However, the body, when dried and ground into powder, was useful for healing wounds. When the Malalis had insomnia, they swallowed the worm powder—created from a worm without the head but with the intestinal tube—and then they fell into an opiumlike "ecstatic sleep," during which they had marvelous dreams. Sometimes the "sleep of the worms" lasted an entire day. Later experiments revealed that the worm's psychedelic properties resided in its intestinal tube and perhaps the salivary glands.

Sadly, the Malalis became extinct long ago as a result of encroaching Western civilization, so we can't ask them which particular species they consumed and the degree to which the worm may have been a portal to new realities or ancient dreams. The Malalis population, also decimated for years from attacks by the neighboring Botocudo tribe, had dwindled to a dozen by the early 1850s.

These worm intestines provide one of the earliest examples of a known hallucinogen coming from insects. I suppose that the U.S. government would make it illegal to eat the psychedelic digestive tracts if the practice became widespread. Drug laws are peculiar and interesting to contemplate. Might it be legal to *possess* the bamboo worms though not to consume them for the purpose of psychedelic research? These days, members of certain Native American tribes in California swallow live ants to induce visions. "The ants bite the stomach lining, injecting their venom, and later may be vomited up, still alive."[7]

Many other natural psychedelics exist. For example, San Pedro cacti contain mescaline, and these cacti may also be found in many plant nurseries and often at your local Home Depot or supermarket. Several species of commercially available ornamental grasses contain the psychedelic DMT. A species of rhododendron, the lavender *ponticum*, is poisonous but also known to induce visions when its smoke is inhaled.[8] All of these plants are legal to possess, though not to consume. The botanical psychedelic iboga comes from an African bush that grows edible orange fruits the size of olives. In particular, iboga comes from the root bark of the bush, and I will discuss the mysteries of iboga in coming chapters.

Dream Fish of Norfolk Island

Even more interesting than the psychedelic worms are the "dream fish" of Norfolk Island. Eat the dream fish—certain species of *Kyphosus* (*Kyphosus fuscus* or *Kyphosus vaigiensis*)—and you will have strange, exquisite dreams.[9] German anthropologist Christian Ratsch, an authority on natural psychedelics, has suggested that the dream fish contain large amounts of the hallucinogen DMT or a related molecule. Whatever the precise molecule, the CIA once feared that the Russians would make use of the dream fish for developing dangerous nerve agents.

A few reporters have eaten the dream fish and described their strange effects. The most famous user is Joe Roberts, a photographer for the *National Geographic* magazine. He broiled the dream fish in 1960. After eating the delicacy, he experienced intense hallucinations with a science-fiction theme that included futuristic vehicles, images of space exploration, and monuments marking humanity's first trips into space.

Perhaps someday I'll visit Norfolk Island to see the dream fish in person. My travel agent tells me that the island is serviced by two airlines, Norfolk Jet and Air New Zealand, with a travel time of two hours from Sydney. The next time you have a map handy, take a peek. The island is located in the South Pacific Ocean, east of Australia, and has a population of about 1,800. British attempts at establishing the island as a penal colony in the 19th century were ultimately abandoned. Most of the coastline consists of nearly inaccessible cliffs.

Ayahuasca Introduction

Christian Ratsch is a colorful character, with waist-length hair and long beard. He and others suggest that "enlightenment" does not require years of meditation but rather can be experienced through relevant drug interactions taking place in the brain.[10] One of the popular routes to enlightenment for many explorers of the mind is through ayahuasca, an Amazonian plant potion containing DMT. We will discuss ayahuasca (pronounced I-yuh-WUHS-kuh) extensively in future chapters. This strong liquid brew is also known as yagé (pronounced yah-hey). In the 1950s, William Burroughs, author of *Naked Lunch*, sought the psychedelic in South America, and he later wrote, "There is nothing to fear. Your ayahuasca consciousness is more valid than normal consciousness...." While in this altered state, he saw "larval beings" that passed before his eyes "in a blue haze."

Sometimes, the visions people experience when taking psychedelics like DMT seem so real, detailed, and with recurrent themes, that I wonder if someday we'll use drugs and brain stimulation to access other realities as valid as our ordinary one. I know this sounds far fetched, but I believe this old idea needs further study. In my novels, I often wonder how humans can best cope with and explore realities separated from ours by thin veils. For example, in *The Lobotomy Club,* scientists and amateurs perform brain surgery on themselves in order to see religious visions and a "truer" reality. Alas, in my book, the new reality turns into a nightmare filled with military conspiracies, insectile alien saviors, bioengineering, Biblical imagery, and prodromic dreams. *The Lobotomy Club* is part of a book series called the "Neorealtity Series" in which readers encounter women and their surgically altered brains, fractal sex, Noah's Ark, hyperspace physics, hallucinating androids, prophetic ants, vitamin B-12, cosmic wormholes, novel plastics, intelligent spiders, and quests for God and the structure of ultimate reality.

The Instinct to Alter Consciousness

Ethnobotanist and ethnomycologist Giorgio Samorni has studied psychoactive substances for decades and believes that the desire to take drugs results from a universal, biologically based urge to expand our minds. His book *Animals and Psychedelics* suggests that all creatures seek transcendence and altered states—from caffeine-dependent goats to nectar-addicted ants, mushroom-loving reindeer, drunken elephants, and intoxicated birds. Psychedelics create new patterns of behavior and alter evolution itself.

For example, Samorni's premise is that many creatures intentionally pursue intoxication. The altered states produced by psychoactive plants "can allow rigid instincts to be bypassed, enabling new behaviors and techniques to be learned and passed along by the experimentalists of a species."[11] Any behavior that increases mating, such as eating sexual stimulants, tends to produce progeny with an inclination to seek these stimulants. Even gorillas and chimps use plants recreationally. Gorillas, for example, know to dig roots of the psychedelic iboga bush to get high. Certain beetles produce abdominal secretions that are so relished by ants that the ants will protect the beetle at the expense of their own larvae. Deer continually search for and consume psychoactive lichen. If the rest of the animal world seeks altered states, could it be only natural that humans seek this as well?

Parallel Worlds

For those of you interested in stories involving nearby realities, go to your video store and check out the 1986 movie *From Beyond*, in which scientists stimulate the pineal gland and open up a door to a parallel and hostile universe. The movie is based on "From Beyond," a story by H. P. Lovecraft (1890–1937).

I'm currently quite addicted to H. P. Lovecraft's stories. He received no serious attention during his lifetime, yet today he has a cult following.[12] Many people around the world consider his tales of cosmic fungi, ancient Gods, and mysterious dreams to be works of genius, and they carefully study his every word. In the movie version of "From Beyond," Doctor Pretorius encounters fishlike life forms that float through our world. We are not aware of these beings due to our limited sensorium, but brain stimulation can awaken humanity to see the new reality.

Lovecraft's original "From Beyond" was written in 1920 and published in 1934. In this haunting tale of parallel universes, before modern
physicists were taking the idea as seriously as they do today, Lovecraft
opens readers' minds to other worlds:

> What do we know of the world and the universe about us? Our means
> of receiving impressions are absurdly few, and our notions of
> surrounding objects infinitely narrow....With five feeble senses, we
> pretend to comprehend the boundlessly complex cosmos, yet other
> beings with wider, stronger, or different range of senses might not
> only see very differently the things we see, but might see and study
> whole worlds of matter, energy, and life which lie close at hand yet
> can never be detected with the senses we have...[13]

The scientist in Lovecraft's story then goes on to describe how his invention will enlarge humanity's perceptions:

> We shall see that at which dogs howl in the dark, and that at which
> cats prick up their ears after midnight. We shall see these things, and
> other things which no breathing creature has yet seen. We shall
> overleap time, space, and dimensions, and without bodily motion
> peer to the bottom of creation.[14]

Lovecraft's story impressed me as a teenager and stimulated my early
interest in the possibility of parallel or "nearby" universes, a subject that
I touch in many of my science books. Lovecraft's protagonist uses an
impressive electrical device to open a portal to a world occupying the
same space as ours and filled with "inky, jellyfish monstrosities which
flabbily quiver in harmony with the vibrations from the machine." The
creatures are present "in loathsome profusion." Dr. Pretorius sees that
"they were semi-fluid and capable of passing through one another and
through what we know as solids."

We'll discuss scientists' view of parallel universes further in Chapter
6, but for now, notice how close Lovecraft's theories are to those of William James, who wrote in *The Varieties of Religious Experience*:

> Our normal waking consciousness is but one special type of
> consciousness, whilst all about it, parted from it by the filmiest of

screens, there lie potential forms of consciousness entirely different. No account of the universe in its totality can be final which leaves these other forms of consciousness quite disregarded. They may determine attitudes though they cannot furnish formulas, and open a region though they fail to give a map.[15]

Science, Science Fiction, Asimov, Sagan

Science fiction, like Lovecraft's "From Beyond," has had a profound effect on the evolution of scientific research and the development of scientific theory. Moreover, science fiction has often been a useful source of scientific ideas. I like to think of science fiction as the "literature of edges" because the topics are poised on the edge of what is and what might be. Certainly, science fiction is a literature of change. Moreover, our universe is a science-fiction universe, filled with mystery—constantly fluctuating and evolving. Isaac Asimov said that "science fiction is the only form of literature that consistently considers the nature of the changes that face us, the possible consequences, and the possible solutions." Most scientists grew up reading science fiction, so how could science fiction not affect scientific research and theories? As one example, young men like German-born Wernher von Braun and Willy Ley were so impressed by Kurd Lasswitz's novel *Two Planets* (1897) that they later became eminent rocket scientists. *Two Planets* (or *Auf zwei Planeten*) was such a rage in Germany that clubs formed to discuss the book, which sold over 200,000 copies in Germany alone.

Two Planets described a fictional Martian society with advanced technology. The humanoid Martians come to Earth to educate humanity, to assist in the development of a utopian Earth. It is not an exaggeration to say that this book had an indirect yet tremendous influence in the development of our space age. Similarly, in Russia, Aleksandr Bogdanov's *Red Star* (1908) stimulated a generation of budding scientists and engineers. Theoretical physicist Freeman Dyson says that his own involvement in the search for extraterrestrial intelligence resulted directly from his reading Olaf Stapledon's science fiction novel *Star Maker*, first published in 1937.

Many of my own science books include science-fiction storylines to stimulate readers' interest in the serious science. For example, my "non-fiction" books *Black Holes: A Traveler's Guide*, *Time: A Traveler's Guide*, *The Stars of Heaven*, *Surfing Through Hyperspace*, *The Mathematics of*

Oz, *The Loom of God*, and *The Paradox of God* all feature fictional characters who investigate astronomy, physics, mathematics, and religion.

One recent anecdote shows the profound degree to which science fiction affects research. While writing his science-fiction novel *Contact*, Carl Sagan searched for a way a character could travel great distances quickly. At first Sagan considered using a black hole, but he understood that black holes would be impractical. Sagan asked physicist Kip Thorne for suggestions, and Thorne replied with a detailed letter containing many equations showing how a wormhole might be made to traverse interstellar distances. Thorne soon realized that the equations implied that wormholes could also be used by physicists to create a time machine—at least in theory. Thorne's serious scientific papers on time travel and wormholes, stimulated by Carl Sagan's need for a science-fiction plot device, led to important breakthroughs in theoretical physics.

In the movie *Contact*, Dr. Ellie Arroway is scolded and told that her proposal to search for extraterrestrial life seems "less like science and more like science fiction." She responds, "Science fiction. You're right, it's crazy. In fact, it's even worse than that, it's nuts. You wanna hear something really nutty? I heard of a couple guys who wanna build something called an airplane, you know you get people to go in, and fly around like birds. It's ridiculous, right?…Look, all I'm asking is for you to just have the tiniest bit of vision…."[16]

The fictional Ellie Arroway is a metaphor for many of the true-life "mad geniuses" who have had an irreverence toward authority and a striking degree of independence. Many trend-setting scientists experienced both social and professional resistance to their ideas. Electronics whiz Nikola Tesla was often laughed at when he proposed correct ideas. Alexander Fleming's revolutionary discoveries on antibiotics were met with apathy from his colleagues. Niels Bohr's doctoral thesis on the structure of the atom was turned down by his university—and yet the work later won him the Nobel Prize. Louis de Broglie was belittled for thinking matter could be both particles and waves, and later won a Nobel Prize for his theories. Joseph Lister's advocacy of antisepsis was resisted by surgeons. Reaction to Alfred Wegener's theory of continental drift was generally hostile and scathing. Einstein's paper "On a Heuristic Point of View Concerning the Production and Transformation of Light," in which he proposed that electromagnetic radiation consists of tiny packets called "quanta" or "photons," was at first rejected by many physicists, including the great Max Planck, who subsequently confirmed the theory. Einstein later received the Nobel Prize for this work, which contributed to the foundation of quantum theory.

The message of Arroway, Einstein, and perhaps even of the dream fish, is to avoid where the path leads, and to go, instead, where there is no path, and blaze a trail.

"As the island of knowledge grows, the surface that makes contact with mystery expands. When major theories are overturned, what we thought was certain knowledge gives way, and knowledge touches upon mystery differently. This newly uncovered mystery may be humbling and unsettling, but it is the cost of truth. Creative scientists, philosophers, and poets thrive at this shoreline."

—W. Mark Richardson, "A Skeptic's Sense of Wonder"

"The Leather Man from Shrub Oak is the wayfarer who connects, through observation and experience, distinct realms of time and space, different layers of existence, including cultural, linguistic, and chemical realities. The Leather Man had to experience life in a variety of ways in order to survive, and his facility for handling disparate realities not only imposed a series of grand challenges but also greatly enriched the Leather Man's existence. The Leather Man is in effect the author, and the author's proxy, the reader."

—Jamie Forbes, personal communication

"Since current drug laws target analogues of illegal substances, this means that any chemical in our bodies that is molecularly similar to a scheduled drug is outlawed. In other words, each one of us—infants, gray-haired grannies, the President—has illegal substances coursing through our veins every minute of the day."

—Russ Kick, *The Disinformation Book of Lists*

"My immediate response to sushi is as a vision of glittering food jewelry—and jewels remind me of the near-universal sumptuary value we place upon shining things, across culture and time. Perhaps it was Aldous Huxley who proposed that the glimmering objects of the material world are reminiscent of the heightened sense of light and color experienced in altered mental states, and this jewel-like world has become a symbol of transcendence whenever encountered."

—Jamie Forbes, personal communication

"In 1945, there were 20,000 Hasidim in the world. Today, there are between 350,000 and 400,000, about half of whom live in Israel. This population explosion cannot be explained simply by demographic reasons."

—Publisher's blurb for Jacques Gutwirth,
La Renaissance du Hassidisme

Scientist Carl Sagan sometimes used marijuana to produce highly acclaimed scientific papers. Sagan writes: "The devastating insights achieved when high are real insights...The illegality of cannabis [marijuana] is outrageous, an impediment to full utilization of a drug which helps produce the serenity and insight so desperately needed in this increasingly mad and dangerous world."

—Carl Sagan, in Lester Grinspoon's
Marijuana Reconsidered[17]

The Quantum Mechanics of Hopi Indians

In which we encounter the Sapir-Whorf Hypothesis, the language of the Tariana, the karass and granfaloon, shibumi, Joumana Medlej, Alexandra Aikhenvald, Kawesqar, secret Hebrew names for God, golems, π: the movie, language as a virus; rabbis Hanania ben Teradion, Judah Low ben Bezalel and Schneur Zalman of Lyadi; beings with five billion mouths, Ös, high-IQ genius Chris Langan, "The Great Eskimo Vocabulary Hoax," time travel, Douglas Adams, Temple Grandin, Freeman Dyson, Jean-Francois Champollion, the Rosetta Stone, Arthur C. Clarke, Kurt Vonnegut, the Tetragrammaton, 666, Lord of the Rings, Kabbalah, free will, William S. Burroughs, Terence McKenna, aboriginal Pintupi, and the Guugu Yimithirr.

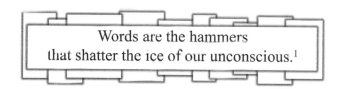

Words are the hammers
that shatter the ice of our unconscious.[1]

Sapir-Whorf Hypothesis?

In the last chapter, I mentioned that my scenic town of Shrub Oak is located in Yorktown, New York. Yorktown comprises seven distinct neighborhoods and five business hamlets: Crompond, Jefferson Valley, Mohegan Lake, Shrub Oak, and Yorktown Heights. The gorgeous Jefferson Valley Mall is just a minute or two away from my home, and one of my science-fiction novels takes place entirely within the confines of this shopping area.

Shrub Oak is in upper Westchester County, about 40 miles north of New York City. European settlement of the Yorktown area began in 1683, when Stephanus Van Cortlandt purchased his first property. Van Cortlandt was a Dutch-American colonial merchant who was also the first native-born mayor of New York City and chief justice of the Supreme Court of New York.

Yorktown actually had some train service in 1881, but by 1958 the trains had stopped running. The old railroad station is now a historical landmark.[2]

As I walk down Main Street, I look from side to side, admiring the oak trees scattered though the rain-dappled village. To my left is a church steeple, jutting through the trees. To my right is a woman in a turtleneck sweater, walking her golden retriever, struggling to keep the dog away from thickets of tulips. My notebook is in my hands.

I think back to a year ago and I remember a walk I took along Main Street while eating a corned beef sandwich purchased at Joe Camera's deli. The sandwich was not unusual, but just as I finished the sandwich, I saw a young lady with long hair holding a sign. She had a necklace made of flowers. The sign read:

Down with the Sapir-Whorf Hypothesis

I stared at her for a minute. I wasn't quite sure what the *Sapir-Whorf Hypothesis* was, but I knew it had something to do with the limitations of language and how language shapes our perception of reality. As wind blew from the north, I jotted down the impressive-sounding words "Sapir-Whorf Hypothesis." The girl smiled at me. I vowed to learn more.

As we muse about language, let's start with the words of one of my heroes, visionary physicist Freeman Dyson, who reminds us that humans are a "species of ape that only recently climbed down from the trees." He writes:

> All our understanding of nature is based on human language. And human language is a tool contingent on the particular history of our species. It would be amazing if human language could comprehend aspects of the universe that no human has seen or experienced. If there are minds in the universe larger than ours, it is unlikely that our language could encompass their thinking. It is unlikely that our science could explain their concepts.[3]

To get a better handle on language and reality, I recently chatted with Lebanese journalist, designer, and Kung Fu aficionado Joumana Medlej— a young woman who shares my curiosity about how words may shape our perceptions. Incidentally, Joumana's first name is derived from the ancient Persian word *juman* (zhu-MAHN) for a "precious pearl." Medlej is an Arabic surname meaning "he who travels by night." Put together, her full name can be translated as: "a night-traveler's pearl."

While debating the finer points of language and reality, Joumana gave me several examples of how different languages compartmentalize the world.[4] For example, English words for the following berries actually end with the word "berry": mulberry, blueberry, raspberry, and strawberry. The French, on the other hand, use entirely different words for the same fruits: *mûre, myrtille, framboise*, and *fraise*. Joumana contends that an English speaker and a French speaker won't be *thinking* of the group in quite the same way. "In the one case," Joumana says, "the thinking process involves a whole that is broken down; in the other, they are separate elements that only come together in the presence of knowledge beyond the words themselves."[5]

"Give me other examples," I said, admiring her curly brown hair.

She proceeded to tell me that the Japanese language doesn't have a single word for water but rather two words: *mizu* (cold water) and *oyu* (hot water). *Mizu* can sometimes be used more generically, but *oyu* can never be cold, although it can be tepid. The point is that neither of these words translates accurately the notion of plain "water." To a Japanese, the two concepts cannot easily be grouped under a single word, as they are in English. I nod, explaining that the Inuit have dozens of words for snow, *if* you count words with multiple suffixes as different words. Michael Fortescue's 1984 book *West Greenlandic* lists over forty "words" for ice and snow, such as: sea ice (*siku*), large expanses of ice in motion (*sikursuit*), "new ice" (*sikuliaq*), melting ice (*sikurluk*), chunks of freshwater ice (*nilak*), lumps of ice stranded on the beach (*issinnirit*), snow blown into a doorway (*sullarniq*), snow falling in air (*qaniit*), hard grains of snow (*nittaalaaqqat*), and feathery clumps of falling snow (*qanipalaat*). West Greenlandic often jams many English words into one compact unit. Perhaps it is not correct to count each of the forty words separately, but it does appear that West Greenlandic has many more compact ways of signifying snow than does English.

Today, controversies rage as to whether the Inuit *really* have richer vocabularies for snow than do English speakers. Many English snow-related words do indeed exist, like sleet, slush, blizzard, and others that are not as commonly spoken. Psychologist Steven Pinker calls the idea that the Inuit have more words for snow than English the "Great Eskimo Vocabulary Hoax."

One of my favorite analyses on the subject is Anthony C. Woodbury's paper, "Counting Eskimo Words for Snow: A Citizen's Guide."[6] Professor Woodbury is at the Department of Linguistics at the University of Texas. In the paper, Anthony attempts to list all "lexemes" referring to snow in the Eskimo language called Central Alaskan Yupik (or simply "Yupik Eskimo"), which is spoken by about 13,000 people in the coast and river areas of Southwestern Alaska. (Anthony notes that, whereas Canadians prefer the term "Inuit" to Eskimo, he uses the term because it properly refers to any Eskimo group, not only the Inuit, and its use is widespread in Native communities in Alaska.)

Yupik Eskimo is one of five Eskimo languages, the most famous of which is Inuit, spoken in various dialects in Alaska, Canada, and Greenland. Anthony defines lexemes as independent vocabulary items or dictionary entries. Thus, English has a single lexeme, "speak," that has several inflected forms, such as speaks, spoke, and spoken. Anthony counts only about fifteen basic Eskimo lexemes for snow, restricting himself to lexemes used in a small subpart of the Central Alaskan Yupik-speaking region. However, he does note that the terms are inflectionally so complicated that each single noun lexeme may have about 280 distinct inflected forms—and each verb lexeme may have more than 1,000! Here are just a few distinct lexemes, to give you a flavor of the language: snowflake (*qanuk),* frost (*kaneq*), fine snow/rain particles (*kanevvluk*), drifting particles (*natquik*), clinging particles (*nevluk*), fallen snow (*aniu*), soft, deep-fallen snow on the ground (*muruaneq*), crust on fallen snow (*qetrar),* fresh-fallen snow on the ground (*nutaryuk*), fallen snow floating on water (*qanisqineq*), snowbank (*qengaruk*), snow carved in a block (*utvak*), snow cornice formation about to collapse (*navcaq*), blizzard snowstorm (*pirta*), severe blizzard (*cellallir*). So the Eskimos have the compact *navcaq* and we have "snow cornice formation about to collapse."

Other cultures certainly have many words for elements of their environment that are important to them. For example, the Tariana of Brazil have numerous compact terms for ants—edible ants, biting ants, stinging ants, and so forth—words for which a "regular" speaker of English would probably not have. However, unlike English, the Tariana do not distinguish linguistically between green and blue.

Language, Color, and the Guugu Yimithirr

Let's talk a bit about color. Dr. Paul Kay, emeritus professor of linguistics at the University of California at Berkeley, has noted that English has eleven basic words for color: black, white, red, green, yellow, blue, brown, pink, orange, purple, and gray. Russian has an additional color, *goluboy* (light blue).[7] The ancient Mayan language of southern Mexico combines green and blue to form what linguists call "grue." This doesn't mean that Russians can see colors that Americans cannot, but our languages affect how we categorize the world and put objects into groups.

Most humans have very similar color categories because we have similar visual systems and live in a world bathed in the same distribution of visible wavelengths. On the other hand, the New Guinea language Dani has just two words for colors! One word covers "hot" colors like white, red, and yellow. The other word covers "cool" colors like black, green, and blue. Kay has suggested that hunter-gatherers may have needed fewer color words because color data rarely provided much "crucially distinguishing information about a natural object or scene."[8] This is difficult for me to understand, because hunter-gatherers see such a plethora of colors in flowers, roots, and insects, and it would seem useful to be able to distinguish subtleties linguistically in order to help with survival. So I remain mystified as to why certain peoples such as the Dani lack words for certain concepts. Perhaps color may become less important within regions of rainforest with a thick tree canopy and frequent cloud cover.

Even more linguistically constrained than the Dani are the Piraha tribe of the Amazon. According to Daniel Everett, an American linguistic anthropologist, the Piraha are the only people known to have no distinct words for colors. Moreover, these hunter-gatherers are the only group of people known to have no concept of numbering and counting. Adult Piraha apparently are unable to learn to count or understand the concept of numbers or numerals, even when the Piraha asked scientists to teach them and even after having received basic math lessons for weeks. This complete lack of number concept has led scientists to describe the Piraha as "something from Mars."[9]

The Piraha have no written language and no collective memory going back more than two generations. They do not sleep for more than two hours at a time during the night or day.[10] They frequently change their names, because they believe spirits periodically possess them and change who they are. They have no creation myths.

Columbia University cognitive psychologist Peter Gordon found that the Piraha could not perform any mathematical operations. For example, when presented with a very small set of objects and asked to duplicate the number they saw, the Piraha could not get beyond two or three before starting to make mistakes. Gordon believes that a major part of the difficulty relates to their number-impoverished vocabulary. Although the Piraha are capable of saying one word to indicate a single object and another for two objects, these words do not mean "one" or "two" in any usual sense but rather a vague sense of "one-ish" (a relatively small amount) or "two-ish" (a relatively bigger amount). The words are the same syllable, pronounced with a falling or rising inflection.

When the Piraha use the one-ish word to discuss something such as a stone, it is not possible to discern whether they are describing a single stone, a small stone, or two stones, because the two-ish word can mean "not many." Professor Gordon says that his constellation of findings is strong evidence for the notion that language determines the nature and content of how we think. Professor Gordon asked himself about the Piraha, "Is there any case where not having words for something doesn't allow you to think about it?" His response: "I think this is a case for just that."[11] Chimpanzees can count up to nine; however, the Piraha can be inaccurate even when they use their fingers to show numbers lower than five.

According to Gordon, "Whorf says that language divides the world into different categories. Whether one language chooses to distinguish one thing versus another affects how an individual perceives reality."[12]

The Piraha language has no word for "number," and pronouns do not designate number. For example, "he" and "they" are the same word. Most standard quantifiers like "more," "several," "all," and "each" do not exist.[13] Piraha children can easily learn number words in Portuguese; however, it appears that adults have lost the ability to comprehend such words, having never been exposed to words for numbers while growing up.

Paul Kay, the linguist we discussed who is interested in color and language, agrees with me that language shapes the way we compartmentalize reality: "There is a wealth of evidence showing that what people treat as the same or different depends on what languages they speak."[14] Indeed, speakers of Guugu Yimithirr, a language of Australia, have a very different concept of "left" and "right" than do Western speakers, and this appears to mold both their thoughts and behaviors in fascinating ways. For example, the Guugu Yimithirr don't even have words like left and right. Its speakers always describe locations and directions using the Guugu Yimithirr words for north, south, east, and west. Thus, they would never say that a tree is on their left, but rather that the tree is west of their position.

Our discussion on language and vocabulary interacting with reality reminds me of my own "mind alteration" when studying entomology and learning all the insect orders (like Odonata, Coleoptera, and Hymenoptera) and external body parts (like thoracic spiracles, tympanum, and pronotum). When I had names for everything, I perceived insects *so* differently, remembered insects so differently, and communicated about insects so differently. The names helped focus and consolidate my attention in strikingly new ways. Certainly, before I knew the names, I could see that one bug had large wings and another did not, or one had a hypognathous jaw and another did not, but I doubt that my mind was tuned to manage, compartmentalize, and take note of this new information. Before I knew the words, did I wonder why the insects were different or did I just accept that a bug is a bug?

Beings with Five Billion Mouths

People who know numerous languages amaze me. Joumana believes that "the more languages one is familiar with, the more subtle his or her thinking will be." If this is true, another of my heroes, Elihu Burritt (1811–1879), had very subtle thoughts, because he was the world's greatest self-instructed linguist! A blacksmith by trade, he studied mathematics, languages, and geography and became known as "the learned blacksmith." He taught himself more than forty languages and translated Longfellow's poems into Sanskrit. Burritt promoted the causes of antislavery, temperance, and self-education, and he was especially interested in ways to establish world peace.

Another fellow fascinated by language was Jean-Francois Champollion (1790–1832). As a child, Champollion became obsessed with studying ancient languages like Hebrew, Arabic, Syriac, Chaldean, Chinese, Coptic, Ethiopic, Sanskrit, Pahlevi, and Persian. He created a 2,000-page dictionary of Coptic, an ancient Egyptian language. "Every day my Coptic dictionary is growing fatter," he said. "The author, meanwhile, is getting thinner."

Champollion went on to decode hieroglyphics by studying the Rosetta Stone, originally discovered in Egypt by Napoleon's troops in 1799. After solving this fantastic mystery, he collapsed into a week-long coma, and I still don't know the reason for his coma. In particular, after he made his discovery, he rushed to his brother and exclaimed, "*Je tiens l'affaire!*" (I have it!), and then dropped to the floor.

The world's greatest living polyglot is Ziad Fazah, who speaks over 55 languages. Fazah was born in Liberia in 1954, but while still an infant moved with his Lebanese parents to Beirut. Today, Fazah continues to learn new languages—his latest acquisition being Papiamento, a Dutch, Portuguese and Spanish blend spoken in the Caribbean islands of Aruba and Curaçao. Mandarin was the hardest language for Fazah to learn because of the large number of ideograms. Fazah hopes that within the next 10 years he can learn the remainder of the world's roughly 4,000 to 6,000 languages. Over 99 percent of these languages belong to a mere 19 language families. Interestingly, about 1,000 of all the languages on Earth are spoken on the island of New Guinea.

Multilingual people like Fazah, however, pale in comparison to the beings in literature with great linguistic skills. My favorite creatures are those fluent in languages too many to contemplate. In particular, according to Islam, the seventh level of heaven, ruled by Abraham, consists entirely of divine light. Each strange being there is bigger than the Earth and can speak more than 31,000 trillion languages![15] If we believe that languages affect our understanding of reality, each of these beings must have an awesome perspective on our daily lives. According to *Brewer's Dictionary of Phrase and Fable*:

> When Mohammed was transported to heaven, he says: "I saw there an angel, the most gigantic of all created beings. It had 70,000 heads, each had 70,000 faces, each face had 70,000 mouths, each mouth had 70,000 tongues, and each tongue spoke 70,000 languages; all were employed in singing God's praises." This would make more than 31,000 trillion languages, and nearly five billion mouths.[16]

Linguistic Holes

Today, I find that many immigrants to America continue to speak their native language at home, but frequently their brains borrow words from English and other languages when they don't have the words in their own.

Herr Hill, my German teacher at Ocean Township High School, told me that the German language has no word for "fair." Similarly, the Breslau historian Hermann Cohn wrote in a 1943 antifascist pamphlet: "The Germans do not even have a word for 'fair,' just as they have no word for 'gentleman'…they are a nation without hobbies." What my teacher did not tell me is that German has fairly close equivalents: "*unparteiisch*" means "impartial," and "*anständig*" is equitable. So perhaps my teacher meant that German does not have a single word that covers all the different meanings of "fair" in English. Anecdotes like this make me skeptical whenever someone tells me that a language does not have a word for a particular concept, because when I speak to a person fluent in the language, they always seem to supply the missing word. I would be delighted to hear from readers who maintain lists of these linguistic "holes" in which one language misses a word for a basic concept that exists in another language.

Sapir-Whorf Hypothesis, Einstein, and the Material Universe

The argument that language partially shapes the way we perceive reality is an old one that gained accelerating interest in the early 1900s when anthropologist and linguist Edward Sapir (1884–1939) proposed that language and thoughts are interconnected like threads in a complex braid, and that humans are often restricted by their vocabularies and languages.

Sapir's student Benjamin Whorf (1897–1941) extended this concept of linguistic reality-shaping and believed that different world views are shaped by different languages. He also suggested that what we actually *think* is determined to a large extent by our language. However, I don't think Whorf would have said that *all* our thoughts are confined to language. For example, you can imagine the visual, olfactory, or aural aspects of an object without resorting to language. You can imagine the face of Britney Spears, the sound of the Beatles, the shape of Saturn, or the smell of roses without words. Even Albert Einstein didn't always resort to words when he thought about the material universe. He wrote in his *Autobiographical Notes* that he often thought in terms of images, and that verbal thinking sometimes occurred only at the end of a nonverbal progression. "I rarely think in words at all," he said. "A thought comes, and I may try to express it in words *afterwards*."[17] Of course, in order to communicate his ideas to others, Einstein had to render his nonverbal thoughts as language. I wish that Whorf could have interviewed Einstein about Einstein's nonverbal thinking.

Whorf was particularly excited about the Hopi language, and he concluded that Hopi speakers do not include tense in their sentences in the same way that English speakers do. Thus, he reasoned that the Hopi people must have a different sense of time from many other people. In English, the form of a verb can determine whether the verb describes a past or present event; for example, "Cliff paints a door" versus "Cliff painted a door." Rather than focus on present and past, Hopi verb forms focus on how the speaker came to know the information. Thus the Hopi would use different forms for firsthand knowledge, such as "I like peanuts," and information that is more general, like "Snow is white." Of course, English speakers may choose to specifically indicate this information, such as in "I was told that Britney wears short skirts," but we usually don't *insist* on this sort of specificity. Given this basic difference, Whorf suggested that Hopi speakers and English speakers think about the world differently, with Hopi speakers focusing more on *how* information about an event is known to be true and English speakers focusing more on the time of the event.

Joumana Medlej, launching from Whorf's ideas, explains that the Hopi consider units of time like pearls in a necklace: "Instead of minutes ticking away, picture the minutes being strung so that every 'present' moment is a string of all the minutes that have passed already."[18] Whorf believed that the Hopi language (and hence culture and thought processes) would potentially allow the Hopi to contemplate certain aspects of quantum physics, or Einstein's theories of space-time, better than people speaking European languages.[19]

One of the most ardent critics of Whorfianism is the psychologist and linguist Steven Pinker, who wrote in his book *The Language Instinct*, "No one is really sure how Whorf came up with his outlandish claims, but his limited, badly analyzed sample of Hopi speech and his long-term leanings towards mysticism must have helped."[20] Pinker also tries to debunk Whorf's claims about time in the Hopi language by referring to anthropologist Ekkehart Malotki's work that seems to suggest that the Hopi have a concept of time similar to ours.[21] However, others like Professor Dan Moonhawk Alford have questioned Malotki's research, and say that it should not be heavily relied upon in this particular case.[22] So the battle continues. Isn't it amazing how much controversy the Sapir-Whorf hypothesis has stirred?

Even today, researchers and authors remain divided about whether thought can exist without language. For example, authors Clive Wynne

(*Do Animals Think?*) and Euan Macphail (*The Evolution of Consciousness*) suggest that thought *requires* language and, thus, neither nonhuman animals nor preverbal children are capable of thought. On the other hand, Jerry Fodor (*The Language of Thought*) argues that language is unnecessary for thought but is simply a means of expressing thoughts.[23] Nicola Clayton of Cambridge University agrees that thought does not require language and referred me to people with high-functioning autism (also called Asperger's syndrome) who think in pictures, not words. For example, Dr. Temple Grandin is an autistic person who feels she can "see through a cow's eyes," and is an influential designer of slaughterhouses and livestock-restraint systems. She writes in her book *Thinking in Pictures*:

> I think in pictures. Words are like a second language to me. I translate both spoken and written words into full-color movies, complete with sound, which run like a VCR tape in my head. When somebody speaks to me, his words are instantly translated into pictures....Visual thinking has enabled me to build entire systems in my imagination...Before I attempt any construction, I test-run the equipment in my imagination....I create new images all the time by taking many little parts of images I have in the video library in my imagination and piece them together. I have video memories of every item I've ever worked with...[24]

What Is Time?

Although the Hopi do not usually include references to the past, present, or future in their grammars, they include two other tenses, *manifested* (physically) and *becoming manifested* (not perceived by the senses). Similar to what we just discussed, *manifested* includes everything in the physical world, past and present but not future. *Becoming manifested* (or simply *manifesting*) includes mental constructs and aspects of reality that cannot be perceived with the senses. Verbs are always expressed within terms of the manifested and becoming-manifested tenses.

Science-fiction authors sometimes resort to exotic tenses when dealing with time travel. Douglas Adams, in *The Restaurant at the End of the Universe*, creates a number of new tense suffixes to represent events that happen as a result of time travel:

The major problem [of time travel] is quite simply one of grammar, and the main work to consult in this matter is Dr. Dan Streetmentioner's *Time Traveler's Handbook of 1001 Tense Formations*. It will tell you for instance how to describe something that was about to happen to you in the past before you avoided it by time-jumping forward two days in order to avoid it. The event will be described differently according to whether you are talking about it from the standpoint of your own natural time, from a time in the further future, or a time in the further past and is further complicated by the possibility of conducting conversations whilst you are actually traveling from one time to another....Most readers get as far as the Future Semi-Conditionally Modified Subinverted Plagal Past Subjunctive Intentional before giving up...[25]

Many Earthly cultures have devised interesting words for long periods of time. For example, one of the longest measures of time is the *kalpa* in Hindu chronology, equivalent to 4,320 million years. The kalpa is divided into 14 periods, and the universe is recreated at the end of each of these periods. We are currently in the seventh of these 14 periods of the present kalpa.

The physics of time is a fascinating subject, which we will address further in Chapter 9. Thanks to Einstein, modern physics views time as an extra dimension; thus we live in a universe having three large spatial dimensions and one additional dimension of time. Stop and consider some mystical implications of spacetime. Thomas Aquinas believed God to be outside of spacetime and thus capable of seeing all of the universe's events, past and future, in one blinding instant. An observer existing outside of time, in a region called "hypertime," can see the past and future all at once. Some mystics have suggested that spacetime is like a novel being "read" by the soul, the "soul" being a kind of eye, or observer, that stands outside of spacetime, slowly gazing up along the time axis.

I'm always fascinated by scientific and cultural studies of the nature of time and space. Ever since H. G. Wells' 1895 publication of *The Time Machine*—which described a four-dimensional spacetime with "duration" being a dimension like height, width and thickness—people have been wondering why we can't travel in time as we do in space. If time is like space, then in some sense the past may literally still exist "back there" as surely as New York still exists even after I have left it. If we could travel in time as easily as we do in space, imagine how our lives would be transformed. We would no longer have regrets about past events. Nor would we wonder about "roads not taken." We'd simply go back in time and make other choices and see what happens. If we were unhappy with the results, we'd try again and again. Time travel would allow us to move in the direction of omniscience, omnipresence, and omnipotence— qualities humans have normally attributed to God. In a sense, time travel would allow us all to become God.

Some physicists believe that if we seriously consider time as a fourth dimension, then the past and future have always existed, and that human consciousness, for some unknown reason, perceives the universe one moment at a time, giving rise to the illusion of a continually changing present. As mathematical physicist Herman Weyl once noted, "The objective world simply is; it does not happen. Only to the gaze of my consciousness, crawling upward along the life line of my body, does a section of this world come to life as a fleeting image which continuously changes in time."[26] Perhaps other beings in the universe do not have our perceptual constraints with regard to future and past, just like the creatures in Kurt Vonnegut's *Slaughterhouse 5*. In this book, the aliens see past and present at once and view human beings as "great millipedes with babies' legs at one end and old people's legs at the other." If time is truly a fourth dimension, then time travel always remains a possibility.

Karass and Granfalloon

As we've been discussing, the Sapir-Whorf hypothesis suggests that the structure of a language colors some aspects of perception and the way we approach reality.[27] Language molds a young child so that he or she has a tendency to adopt a certain world-view. Of course, there is a bidirectional memetic flow so that a culture is *mirrored* by its language and language is molded by culture.

Let's hear it in Sapir's own words. In 1928, Sapir wrote in *Language:*

It is quite an illusion to imagine that one adjusts to reality essentially without the use of language and that language is merely an incidental means of solving specific problems of communication or reflection. The "real world" is to a large extent unconsciously built up on the language habits of the group…We see and hear and otherwise experience very largely as we do because the language habits of our community predispose certain choices of interpretation.[28]

Benjamin Whorf wrote:

> We dissect nature along lines laid down by our native languages. The categories and types that we isolate from the world of phenomena we do not find there because they stare every observer in the face; on the contrary, the world is presented in a kaleidoscope flux of impressions that has to be organized by our minds—and this means largely by the linguistic system in our minds. We cut nature up, organize it into concepts, and ascribe significances as we do, largely because we are parties to an [unstated] agreement to organize it in this way.[29]

Whorf suggests that we should consider the notion of linguistic sculpting of perception because "no individual is free to describe nature with absolute impartiality but is constrained to certain modes of interpretation even while he thinks himself most free."

Some of my colleagues frown when I claim that language can shape the architecture of our thoughts. But after repeating a Baha'i prayer, reading about the word "grok" in Heinlein's *Stranger in a Strange Land*, or "shibumi" in Trevanian's *Shibumi*, I perceived the universe in a whole new light, and it's not an exaggeration to say that compact expressions that consolidate a complex set of concepts have changed my life. There are numerous examples of these supramemetic words from literature—karass and granfalloon from Vonnegut are just two others that come to mind. In social structures, the *granfalloon* is a group of people united or organized by decree and official hierarchies, a bureaucratic structure. A granfalloon within a corporation may often be constrained and ineffective. On the other hand, the *karass* are those social networks that *actually get the work done.* Outside a company, a karass is a spontaneously forming group, joined by unpredictable or informal links. I love working in the karass. My own karass of friends forms a hive mind, a group brain that I find immensely useful.

Some readers will suggest that it's the *concept* of the word "grok" or "shibumi" that changed me and a single word was not needed, but I don't think that's entirely the case. It's as if I needed a compact word to help me express and remember a complex concept. I needed the word so that my brain could easily wrap itself around the concept and reference it again and again, chanting it like a mantra. For me, a word is a vessel, holding liquid ideas that are apt to flow away if not enclosed. The words function as spoons that we dip into the vast landscape of reality. We pull spoonfuls from the sand and call those spoonfuls the world. In mathematics, this is clearer. If you've never had exposure to the words and symbols of mathematics or certain branches of mathematical physics, it would be very difficult to calculate, communicate, or develop practical applications regarding string theory, infinite-dimensional Banach spaces, or Fourier analysis.

Babel's Children

Someday I hope to journey to the faraway Riau islands that straddle the Strait of Malacca. The territory includes a large slice of the eastern Sumatran coast and more than 3,000 islands of all sizes.

According to David Gil, a researcher at the Max Planck Institute for Evolutionary Anthropology in Leipzig, Riau Indonesian is a language, like many others, that does not distinguish nouns and verbs.[30] For example, the phrase "the man is swimming" might translate to a phrase that means "man swim." But the same pair of words could have many other meanings as diverse as "the man is making somebody swim" or "somebody is swimming where the man is." According to Gil, there are no modifiers that distinguish the tenses of verbs. Gil writes, "Though Whorf's hypothesis fell into disfavor half a century ago, it is now undergoing something of a revival."[31] This revival is due in part to the work of Lera Boroditsky, a researcher at the Massachusetts Institute of Technology. Boroditsky conducts experiments she believes provide evidence that the structure of language affects the way people think. For example, Boroditsky shows three pictures to people. Imagine that the pictures show one of two women, Britney or Christina. Subjects are then shown pictures:

1. Britney about to kick a ball.
2. Britney just finished kicking the ball.
3. Christina about to kick a ball.

Indonesians choose photos 1 and 2 as being most similar. English speakers choose photos 1 and 3 to be the most similar, which suggests that English speakers emphasize temporal aspects of a scene rather than relationships between the objects. Dr. Gil, who works with Boroditsky, says this difference might be due to English requiring every verb to have a tense (such as past, present, future). In Indonesian, expressing a verb's tense is optional.

Alexandra Aikhenvald, Tariana

One of my favorite living linguists is Alexandra Aikhenvald, Professor of Linguistics and Deputy Director of the Research Centre for Linguistic Typology at La Trobe University in Melbourne, Australia. Aikhenvald is an expert on Tariana, an Amazonian language in which it is a grammatical error to report something without saying *how* you determined the information—reminiscent of Hopi verb constructs. Imagine the degree to which the Tariana language could influence politics, police work, and the court system. For example, in English, a teacher can tell a student: "Today I talked to a famous scientist," and the student will not ask: "How do you know you talked to a scientist?" However, in Tariana, the speaker adds to the end of a verb a suffix that informs the listener how the speaker knows something. Thus, in Tariana, a person might say something like "I talked-(face-to-face) to the scientist." Alternatively, the speaker might say, "I talked-(non-face-to-face) to the scientist," if the teacher communicated via a letter, phone, or Internet.

In America, these constructs could be useful in the court. Imagine if a witness automatically said *how* they knew about certain facts in a case. The language might require fewer words on the part of our lawyers and judges or make it easier to prosecute perjury! Alas, thc Tariana language is about to become extinct. In light of such disappearances, we risk losing insights into different ways of thinking or categorizing knowledge.[32]

Alexandra required ten years to fully understand the grammar of Tariana! She believes that languages teach us how the mind works by showcasing categories that are sufficiently important for people to express in words. If these exotic languages die, humanity will be left with fewer world views and fewer examples of ways to perceive and catalog reality.

Kawesqar, Ös

Only six speakers of the Kawesqar (Ka-WES-kar) language exist on a remote island in Patagonia—southern South America between the Andes and the South Atlantic. Kawesqar is the language native to Patagonia since the last ice age! When these six people in the tiny fishing village on Wellington Island die, the language and reality view die with them. According to *New York Times* writer Jack Hitt, half of the more than 6,000 languages currently spoken in the world will be extinct by the end of the century. "In two generations," he says, "a healthy language—even one with hundreds of thousands of speakers—can collapse entirely, sometimes without anyone noticing."[33]

Today, 438 languages have fewer than 50 speakers. "With each language gone, we may lose whatever knowledge and history were locked up in its stories and myths, along with the human consciousness embedded in its grammatical structure and vocabulary...." Hitt suggests that, because of Kawesquar's nomadic history, they rarely used the future tense due to uncertainties in the future. However, the past is rich with memories and legends, and thus the Kawesquar have fine gradations for past tense. For example, "A dog barked" can be linguistically controlled to mean the dog barked a second ago, a day ago, or a time so long ago that the speaker was not the observer of the event (but the speaker personally knew the observer). "A dog barked" can also be controlled to mean that it occurred in a mythological past, a tense so old that it conveys a universal truth.

For a final fascinating example of a dying language, consider Ös. Only a handful of fluent speakers of this Turkic language are still alive in remote Russian villages along the Chulym River in Siberia. Ös has an incredible number of words for fishing equipment and fish. However, today, ethnic Chulym families raise their children to speak Russian. As the Ös speakers shift to speaking Russian, they seem to be losing most of the Ös fishing terms. Linguist K. David Harrison of Swarthmore College, who visits the villagers, believes that "when you lose a language, you also lose specialized knowledge. Each language that vanishes without being documented leaves an enormous gap in our understanding of some of the many complex structures the human mind is capable of producing."[34]

Harrison is also an expert on Tuvan and other Siberian languages. During field expeditions, he often lives with nomadic people, walking or riding with them on their seasonal migrations as they herd camels, horses, yaks, and sheep. In addition, he has worked with one of the last speakers of the Karaj language in Lithuania.

Today, the youngest fluent speaker of the Chulym, a truck driver named Vasya, is in his fifties. Ös was one of many languages that Soviet educators banned when they imposed Russian on local groups in Siberia more than 50 years ago. Harrison feels that, when a language becomes extinct, we lose an opportunity to understand "the variable cognitive structures of which the human brain is capable."[35]

Language as a Skin
Let's wrap up our discussions of the Sapir-Whorf hypothesis by consulting the famous *The Cambridge Encyclopedia of Language*, which

distinguishes between four related hypotheses. The first is that people have thoughts and put them into words. In this sense, language is like a skin for internal thoughts. The second is the Sapir-Whorf hypothesis, which we have discussed: language shapes thought. The third is that language and thought are identical. Here, thought is an internal form of speech. The fourth is that that language and thought are interdependent. This is a widely held notion today. Neither language nor thought takes precedence over the other.

Certainly, Hopi can tell the difference between birds and airplanes, even if the Hopi have one word for "things that fly." Via circumlocution, one can usually describe a thing, concept, or emotion with a set of words if one word is not sufficient. However, as we discussed, an individual may have an easier time expressing a concept if there is a word for it.

Professor David Crystal has devoted his life to the study of language and has published more than a hundred books. In his breathtaking *The Cambridge Encyclopedia of Language*, he presents fascinating information on aboriginal languages.[36] For example, the Australian aborigines must be the world experts on holes because their Pintupi language has ten "words" for different types of holes:[37]

> *yaria*: a hole in an object
> *pirti*: a hole in the ground
> *pimki*: a hole formed by a rock shelf
> *kartalpa*: a small hole in the ground
> *yulpilpa*: a shallow hole in which ants live
> *mutara*: a special hole in a spear
> *pulpa*: a rabbit burrow
> *makampa*: a goanna (lizard) burrow
> *nyarrkalpa*: a burrow for other small animals
> *katarta*: a hole left by a goanna (lizard) when it has broken the surface
> after hibernation

It's true that English also has a lot of words for hole—many almost synonyms, like opening, aperture, crevice, cavity, pit, and so forth—but these words tend to be general and give no information as to the hole's origin. The Pintupi words are not synonyms, and they represent a distinctly aboriginal categorization of reality.

The Parchments Burn But the Words Fly Free

I've always been in love with words, and since high school have filled notebooks with colorful words and phrases that I have encountered while reading. Words are meant to be petted and stroked. They allow us to transcend space and time, and to inspire visions.

Some words, however, should be approached with caution. The ancient Hebrew name for God was thought to be so powerful that merely saying it could get one into deep trouble and cause the universe to rend. One of God's several names is the Tetragrammaton, which we might represent with English characters as YHVH. It's a four-letter name of God first mentioned in the book of Genesis and usually translated as "the Lord." Although there are many names of God in the Old Testament, this one occurs most often (6,823 times) and is written in Hebrew as:

יהוה

This is the name that God gives himself when talking to Moses at Mount Horeb in the Sinai wilderness. Because Jews almost always prohibited its utterance, the actual pronunciation of the Tetragrammaton is not known today. יהוה is commonly pronounced Jehovah, though this is inaccurate. Some scholars conjecture that it was pronounced "Yahweh." The Tetragrammaton may have a cryptic meaning such as "I Am the One Who Is" or "He Causes to Become."

Let's trace this word back in time a few centuries. One of the most important functions that occurred in the Temple of Jerusalem was the pronunciation of God's name. In practice, the Tetragrammaton could only be spoken safely by the High Priest in the Temple because it needed the huge physical and ritual structure to contain its power. Because the Temple was destroyed, orthodox Jews never pronounce God's name for any reason.

In the second century AD, Jewish scholar Hanania ben Teradion was famous for his Bible teachings and for teaching the personal name of God to his students. Alas, in ben Teradion's day, the Roman emperor had made it illegal to teach and study the Torah. Ben Teradion was arrested, wrapped in a Torah scroll, and burned at the stake. The *Encyclopedia Judaica* says: "in order to prolong his agony, tufts of wool soaked in water were placed over his heart so he would not die quickly."[38]

As the fires started, his students asked, "Rebbe, what do you see?" He answered, "The parchments burn, but the words fly free."

They pleaded with him to open his mouth so that the smoke and fire would kill him quickly, but he decided not to take this easier path. "Let the one who gave me life take it away and let me not harm myself."

According to legend, the executioner felt so guilty he asked ben Teradion to request forgiveness for him in the world to come. Ben Teradion agreed, and the executioner removed the sponge and threw more kindling until ben Teradion was burning, and then the executioner himself jumped into the flames.

To increase the horror of ben Teradion's punishment, the Romans executed his wife by the sword and forced his daughter into prostitution. The point of this is that, although the Romans were responsible for Haninah's horrifying ordeal, the Talmud says that ben Teradion had also brought down God's wrath because he had pronounced God's name in public.

Similarly, Rabbi Akiva ben Joseph (50–135 AD) was killed by the Romans because he taught the Torah during a time when such practice was outlawed by the Romans. While the Roman legionnaires tore his flesh from his bones with large iron combs, ben Joseph continued reciting the Shema, a Jewish prayer. Legend has is that ben Joseph's horrified students watched the torture and asked how he could recite the prayer while in such agony. His reply: "All my life I have wondered about the verse in the Shema that says you shall love God with all your mind, strength, and soul. I know how to love God with my mind and my strength, but I could not understand how to love God with my *soul* until now." He repeated the Shema: "Listen Israel! The Lord is God, the Lord is One," and finally died as his voice faded on "One." Ben Joseph is famous for promoting "thou shalt love thy neighbor as thyself" as the primary principle of the Torah.

Bigger Names of God

Kabbalists—Jewish mystics—often meditated on and studied other names of God while seeking to understand and shape reality. Sometimes, Kabbalists inscribed these mysterious names on protective amulets. People who were not properly prepared or who made mistakes in uttering these names of God were said to risk injury or death.

Many of these reality-changing names were top secret. Today, we don't know the meaning of all nor how to pronounce them. We've already mentioned the Tetragrammaton, יהוה, whose pronunciation has been lost.

In the following paragraphs, I will write a few of the secret names, while hoping not to tear the fabric of reality in the process. For example, Eleazar ben Judah of Worms (c. 1165–c. 1230) described a mysterious 22-letter name of God, though its actual origin may be much older. Eleazar was the last major scholar of the German Hasidic movement and saw horrifying persecution of the Jews by the Crusaders who murdered his wife, daughter, and son. The 22-letter name appears in *Sefer Raziel* (c. 1100 CE), an anonymous collection of cosmological and magical works:

אנקתם

פסתם

פספסים

דיונסים

No one knows for sure how it is derived or how to pronounce it. Many Jews I consulted said we should not even try to pronounce it. Various words meaning "miracles" or "their cry" or "suffering" seem to be hiding in this name. Note that the name has 22 letters, and the Hebrew alphabet has 22 letters.

God's four-lettered and 22-lettered names were thought to be powerful, but God also has other mighty names. Some Kabbalists believed that God had a 42-letter name through which human beings could change reality. I risk my soul by writing it here:

אבגיתץ קדעשטן

נגדיכש בטרצתג

בקדטנע יגלפזק

שקוצית

Like the other names, we don't know how to pronounce this one. I've read in a few places that it may be a scrambling of the first 42 letters of the Torah and dates back to the first century. It's mentioned in the Talmud. Line 2 might be interpreted as "Destroy Satan."

I found another 24-letter name of unknown origin, which is printed in the 12-volume *Jewish Encyclopedia* (originally published between 1901 and 1906, and now available at JewishEncyclopedia.com):

יור הא יוד הא ואו הא אלף דלת נון יוד הא ואו יוד הא אלף הא

God also has a 216-letter name composed of groups of three-letter triads. To construct the name, we must examine the words of Exodus 14: 19–21, each of these verses having 72 letters. This name of God is made by joining the first letter of verse 19, the last letter of verse 20, and the first letter of verse of 21. To create the second triad, we use the second letter of verse 19, the second to last of verse 20, and the second of verse 21, and so forth. According to legends, Moses received the 216-letter name of God when he saw the Burning Bush on the mountain and later used it to part the "Red Sea." If one says the name when in "a state of impurity," the word will strike the speaker dead. This complicated name of God, as outlined by the Kabbalists, can be converted into a list of the names of 72 angels seen be Jacob ascending the ladder in Genesis 28: 12. Here are the 72 triads:

כ	ל	ה	ה	מ	ה	י	מ	ה	ל	א	ה	כ	א	ל	מ	ע	ס	י	ו				
ל	א	ק	ר	ב	ז	ה	א	ל	כ	ה	ל	ה	ל	י	ל	ה							
י	ו	מ	י	ה	ל	ע	ו	ד	י	ת	א	ה	ש	מ	ת	י	ו						
מ	כ	ל	י	ל	ו	א	ל	ר	ש	י	ה	ג	ה	מ	י	נ	פ	ל					
נ	ו	ה	ה	ש	כ	ו	י	א	ר	ת	ה	ל	י	ל	ה	ו							
ד	ק	ה	ו	ר	ב	מ	י	ה	ת א	ה	ו	א	י	כ	ל	ו							
ג	נ	ע	ה	ד	ו	מ	ע	ע	ס	י	י	ו	מ	ה	י	ר	ה	א					
י	ג	נ	מ	ה	נ	ה	י	ש	ר	א	ל	ו	י	ה	י	ה	ע	נ					
ת	א	מ	ש	י	ה	ו	ה	ל	י	ל	ה	ל	ז	ה	ו	ע	מ	י					
מ	ה	י	ר	ה	ת	א	מ	ד	מ	ע	י	י	ו	מ	ה	י	נ	פ	מ				
ו	י	ב	א	ב	י	ג	נ	ה	מ	ה	ע	ר	י	מ	ו	ב							
מ	י	מ	ה	ו	ע	ק	ב	י	ו	ת ה	ב	ר	ה	ל	מ	י	ה						

216–Letter Name of God Composed of 72 Triads

For example, the first angel's name is Vehu or וֹ הֹ וֹ, which is the triad in the top row at the right. There are 216 characters in total.

Language Is a Virus

From the thirteenth century onward, Kabbalah generated an extensive literature in parallel with the Talmud—and sometimes in opposition to the Talmud. Most forms of Kabbalah teach that every letter and word of the Torah contains a hidden meaning. Kabbalists suggest that the world is created and sustained by divine language, and that the Hebrew alphabet has a mystical significance. According to Rabbi Yehuda Patiah, a famous Kabbalist of the nineteenth century, hearing one of God's names in any language during a dream means that the dream will come true. In fact, the Talmud states that three types of dreams are likely to be fulfilled: (1) a dream occurring just before waking in the morning; (2) a dream that a friend dreams about you; and (3) A dream interpreted within the dream itself.[39] Additionally, the Talmud implies that dreams need to be interpreted to decipher their meaning; however, the precise interpretation one hears will *actualize* the interpretation, as if the dream represented a set of possibilities and the interpretation tied one of them to reality. Thus, one had to be quite careful to enlist dream interpreters who tended to give positive readings.

Schneur Zalman of Lyadi—founder of the Chabad-Lubavitch Chassidic Movement—suggested that the entire Torah is the name of God. According to Zalman, if God's "letters" were to return to their source, the universe and all creation would never have existed, because their very idea would be nonexistent.

Visionary writer Terence McKenna (1946–2000), who attempted to study reality through the use of psychedelics like DMT, also thought that letters, words, and language were intimately tied to reality. In particular, he thought that reality isn't composed of particles like protons, electrons and quarks but rather of nouns, verbs, and adjectives. He believed that DMT makes syntax visible and lets us visualize the logic behind language and thought, the "machine code of reality." For McKenna, "the universe is made of language." I'll describe McKenna in greater detail in Chapter 3, but what is it that McKenna truly means? I think McKenna is suggesting that language replicates itself through time, creating societies in which both people and machines depend on language of one form or another. He says in *The Archaic Revival*, "Earth is a place where language has literally become alive. Language has infested matter; it is replicating and defining and building itself. And it is in us."[40] In some sense, reality is altered and brought into existence from myths and fables. Similarly, other writers like Grant Morrison believe that magic and language can help us shape existence. He writes in a Kabbalist fashion, "'Magic' is the hopelessly inadequate Standard English word for the long-established technology which permits access to the 'operating codes' underlying the current physical universe."[41]

Chris Langan, one of the smartest people in America—with an IQ off the charts—was once asked if language limits our thought. His response:

> Most people think that a language must be a natural, spoken or written language like English, French or German. But in mathematics and logic, the definition of language is far more general. Reality itself can be viewed as a language, and the language of thought is closely related to that of reality at large. So we might just as well say that the language of thought, rather than limiting our ability to understand reality at large, facilitates it. However, if one's mental language is artificially constrained, then one's insight and creativity may indeed be impaired. The surest proof against this is an open mind and a desire to learn.[42]

Writers like William S. Burroughs and Mark Pesce have suggested that language is actually a virus. Pesce writes:

> Our linguistic abilities aren't innate. They are not encoded in our DNA. Language is more like *E. coli*, the bacteria in our gut, symbiotically helping us to digest our food. Language helps us digest phenomena, allowing us to ruminate on the nature of the world.[43]

Pesce believes that nonhuman animals perceive reality directly, but humans are clouded by the "fog" of language, which invaded and "colonized our cerebrums" millennia ago and thus inserted a layer between us and reality. Pesce writes, "While we think ourselves the masters of language, precisely the opposite is true. Language is the master of us, a tyranny from which no escape can be imagined."

In addition to language mediating our interactions with reality, scientists now know that our voluntary motions are not even initiated by our conscious minds! Brain signals relating to our acts occur *before* we have a conscious intent to do something, such as deciding to push a button. If I had placed an electrode in your brain, I could know when you were about to do something before you knew.[44] Similarly, Pesce says that we arrive at our decisions through emotional sensations, and that we are acting "from the gut at all times." Our "reason" enters the process only *after* the decision has been made.[45]

If Terence Mckenna were alive today, I'm sure he would enjoy talking to Mark Pesce about language. Pesce, one of the coinventors of VRML, a language that brought virtual reality to the World Wide Web, thinks that we will someday apprehend the entire universe as code. A forthcoming "theory of everything" will be a *program*, "a series of linguistic statements, which, like words in a sentence, describe the execution of reality."[46]

Golem Activation via Hebrew

A golem from Hebrew mythology is an animated creature, crafted from clay and whose being is intimately tied to language. Probably the most famous golem is the one that Judah Low ben Bezalel (1511–1609) allegedly created to defend the Prague ghetto from anti-Semitic attacks.

A golem is usually inscribed with magical or religious words that keep it animated. For example, golem creators wrote the name of God on the golem's forehead, or on a clay tablet under its tongue, or wrote the word Emet ("truth" in the Hebrew language) on its forehead. By erasing the first letter in "Emet" to form "Met" ("death" in Hebrew), the golem can be destroyed.

Some ancient Jewish recipes for creating a golem required a person to combine each letter of the Hebrew alphabet with each letter from the Tetragrammaton (YHVH), and pronounce each of the resulting letter pairs with every possible vowel sound. The Tetragrammaton serves as an "activation word" to pierce reality and energize the being.

Most golems in the medieval literature were dumb but could be made to perform simple, repetitive tasks. The challenge for the golem creator was figuring out how to get the golem to *stop* doing the task.

Some Tolkien fans have suggested that the creature named "Gollum" in Tolkien's *Lord of the Rings* derives its name from the word golem. Gollum seems to be a slave to the magical ring in the same way that the golems of Jewish myth were slaves to their creators. Other Tolkien scholars say that Tolkien's Gollum name comes from a swallowing noise the creature makes. In *The Fellowship of the Ring*, Gandalf says, "He took to thieving, and going about muttering to himself, and gurgling in his throat. So they called him *Gollum*, and cursed him, and told him to go far away…" Of course, Tolkien might have used the world for both purposes, to conjure the sound of the word as well as the connotation of golemic slavery.

The word "golem" appears only once in the Bible (Psalms 139: 16), where it refers to an imperfect or unformed body. The New International Version translates the verse as: "Your eyes saw my unformed body. All the days ordained for me were written in your book before one of them came to be." In Hebrew, "golem" signifies a "shapeless mass," and the Talmud uses the word to imply "imperfect."

π: the Movie

Before concluding this chapter on language and reality, note that the mathematician's quest to understand infinity parallels more traditional attempts to touch God through language and prayer. Today we use the symbols of both language and mathematics to express relationships between humans, the universe, and the infinite. Mathematicians and priests seek "ideal," immutable, nonmaterial truths and then venture to apply these truths in the real world.

For a long time, I've been interested in the Jewish use of numbers as a kind of language to access God and reality. This application is exemplified by the 1998 cult movie titled "π," which features as protagonist a mathematical genius who is fascinated by numbers and their diverse role in the cosmos, the stock market, and Jewish mysticism. In the movie, Hasidic Jews interested in Kabbala search for a sequence of 216 numbers that may be represented by Hebrew letters to spell out God's true name, which was destroyed with the Temple by the Romans.

The movie is notable in that, as mentioned above, it's about a mathematical genius named Max who is obsessed with numbers. Max rarely leaves his apartment and sees numbers and number patterns in everything around him. When his computer displays the mystical 216-digit number, it crashes and a strange adventure ensues. Perhaps the number even makes the computer conscious. The audience is never sure. Max decides that the 216-digit number is responsible for his own ill health, and the movie closes as he drills a hole in his own skull as a cure. Weird ending!

Computer guru Michael Egan points out that the movie does not always display the same 216 digits when referring to the mystical number, and sometimes it shows a 218-digit number as the God number:

94143243431512659321054872390486828512913474876027
67195923460238582958304725016523252592969257276553
64363462727184012012643147546329450127847264841075
62234789626728592858295347502772262646456217613984
829519475412398501

The characters in the movie never explain how they arrive at the number of digits. Perhaps these digits correspond to the 216-letter name of God I mentioned previously, composed of letters from the words in Exodus 14:19–21. Notice also that if we multiply the digits in the Number of the Beast, 666 from the book of Revelation, we get: $6 \times 6 \times 6 = 216$.[47]

In Arthur C. Clarke's story "The Nine Billion Names of God," computer scientists team up with Tibetan priests to list all the possible names of God. The computer operators print combinations of letters to produce permutations of names of less than 10 letters. As they complete their printing task, the stars in the sky begin to wink out. God's name has been discovered. Reality fades away like a dream.

A friend at work said it should not be hard to actually pronounce the secret four-letter name of God, the Tetragrammaton יהוה, by having his computer consider every possible vowel, or phoneme, combination with the four consonants. He coupled the output of his computer to a speech synthesizer, so the computer would actually speak the words at a rate of one name per second. At some point, he theorized, the computer should say the name of God. My friend was not concerned about the wrath of God because it was the *computer* uttering the name of God, not him. During one vocalization, my friend's computer crashed. He stopped the experiment and never ran the program again.

"This desire to communicate is basic both to science and to poetry. The scientist seeks to find the order of the universe through the discipline of experiment; the poet, through the discipline of language."

—Helen Poltz, *Imagination's Other Place*

"There is this belief that we perceive the world the way we do because of shared convictions, which are expressed in language…If we could just think about the world differently, we'd look around and see everything in a different way, but we can't do that because language is built from the shared perceptions of the community. [Perhaps] this limitation blocks our perceptions of many things, so that we're prisoners of language. If language influences the way we see, and even limits our view of reality, then language is hugely important. If we restrict our use of it, we're restricting our vision."

—Dean Koontz in Katherine Ramsland's
Dean Koontz: A Writer's Biography

"The fish trap exists because of the fish; once you've gotten the fish, you can forget the trap. The rabbit snare exists because of the rabbit; once you've gotten the rabbit, you can forget the snare. Words exist because of meaning; once you've gotten the meaning, you can forget the words. When can I find a man who has forgotten words so I can have a word with him?"

—Fourth-century Chinese Taoist Chuang Tzu

"Words or language, as they are written or spoken, do not seem to play any role in my mechanism of thought."

—Albert Einstein, quoted in Jacques Hadamard, *An Essay on the Psychology of Invention in the Mathematical Field*

"The universe is a linguistic process. We know that words shape the world as we see it, but now we have come to understand that words shape the world as it is.…The more we learn about how to modify the world, the more that language becomes convergent with reality.…Words are colonizing the world."

—Mark Pesce, "The Executable Dreamtime," in *Book of Lies*

Bertrand Russell's
Twenty Favorite Words

In which we encounter the terraqueous chrysoprase, Bertrand Russell's twenty favorite words, Piers Anthony's ten favorite words, logophiliacs, Oulipoian states of mind, constrained writing, Pynchonomancy, Annie Sprinkle, Robert Sawyer, divination, the mystery of italics, Brion Gysin's Dream Machine, music, painting, plot aesthetics, Dia Center for the Arts, lipograms, The Anagrammed Bible, *Michael Shermer, "A Glass Centipede," computer poetry, creativity machines, Rachter, Ray Kurzweil, Yggdrasill, alien beauty, alien pornography, the Reverend Jerry Falwell, adipocere, Monongahela, Harlan Ellison, Antonin Artaud, Aldus Manutius, the world's largest vocabulary, and amphigory.*

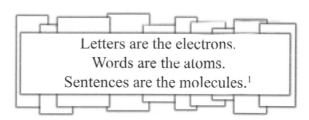

Letters are the electrons.
Words are the atoms.
Sentences are the molecules.[1]

Proust, Einstein, Russell

Marcel Proust's *In Search of Lost Time* is the "greatest and most rewarding novel of the twentieth century, outdistancing its closest rivals, James Joyce's *Ulysses*, Thomas Mann's *The Magic Mountain*, and William Faulkner's *Absalom, Absalom!*"[2] So says Roger Shattuck, author of *Proust's Way: A Field Guide to In Search of Lost Time*. British novelist Graham Greene (1904–1991) said, "Proust was the greatest novelist of the twentieth century." Proust believed that a book was a dynamic entity because a person could read it many times, with new meaning emerging each time the book was read. Thus, rereading a book from one's teenage years can be enlightening because it allows readers to see how they have changed.

One of my favorite books from my teenage years is: *Dear Bertrand Russell: A Selection of His Correspondence with the General Public 1950–1968.*[3] The book contains all sorts of letters from Russell's fans, along with his witty responses. I'll now use Russell's book as a trigger for a zany chapter on people's favorite words and aesthetics in general.

My notebook rests safely in my hand. The air is sweet with the scent of lilacs. I sit on a bench in front of the Hart Library in Shrub Oak. In the last chapter, you learned about my fascination with language and words. Today, I'd like to continue that discussion, and also talk about what you and I consider beautiful—but first a few words about my home town's local library, which I frequently visit to borrow books and audiotapes, learn new words, and generally elevate my mind.

The John C. Hart Memorial Library received its charter from New York State in 1920. Catherine Dresser, daughter of John C. Hart, died in 1916, and in her will she left her family homestead and 45 acres of land to the Town of Yorktown. John C. Hart was born in Shrub Oak in 1822 and attended the district school here. Later he worked in New York City, but his heart and soul were always in Shrub Oak.[4]

Hart's Shrub Oak house underwent frequent remodeling and finally became the Hart Library, in front of which I now sit. Today, nothing of the original 1920 library walls remains. All that exists of Hart himself is an old portrait now hanging on the first floor in the paperback section. He and his wife are buried in the Methodist Churchyard in Shrub Oak. Hart loved his house and the land, and wrote poetry.

Nowadays, the library has a great collection of books by philosopher Bertrand Russell (1872–1970). Russell was many things: a mathematician, a logician, an atheist, a champion of peace, a controversial political figure, and a recipient of the Nobel Prize in literature. Russell so impressed Einstein that in 1931 Einstein wrote to him, "The clarity, certainty, and impartiality you apply to the logical, philosophical, and human issues in your books are unparalleled in our generation."[5]

The Terraqueous Chrysoprase!

In 1958, a fan asked Bertrand Russell to list his twenty favorite words in the English language. Russell replied, "I had never before asked myself such a question," and then proceeded to give his list as follows:

wind	apocalyptic	quagmire	chorasmean
heath	ineluctable	diapason	sublunary
golden	terraqueous	alabaster	alembic
begrime	inspissated	fulminate	chrysoprase
pilgrim	incarnadine	astrolabe	ecstasy

Isn't this a lovely set of luminous words? When examining lists like this, we can compute an "obscurity index" (OI) that tell us the degree to which a person tends to choose words not commonly used by English speakers. When I compute the OI, I simply note the number of words my Lotus WordPro spell-checker does not understand and divide this number by the number of words in a person's list. Russell's obscurity index is 0.25 because my word processor does not understand five of his twenty words (terraqueous, inspissated, incarnadine, chorasmean, chrysoprase). The higher the index, the more "obscure" or "arcane" the list is. For those of you interested in other indices of obscurity, you can simply use Google to count how many Web pages exist that contain the words.

It's obvious I'm in love with these fancy-sounding words. How many of Russell's words do you like or comprehend? I am amazed that such a busy and famous person as Russell would even answer this question about favorite words posed by an unknown fan! Russell received around 100 letters a day. Scholars estimate that Russell wrote one letter every thirty hours of his life. Whenever he changed residences, Russell said that it was "usual for quantities of paper to be burned." When asked how he coped with the onslaught of letters and with writing all his articles and books, he replied by highlighting his daily schedule: "From 8 to 11:30 AM, I deal with my letters and with the newspapers. From 11:30 to 1 PM, I am seeing people. From 2 to 4 PM, I read, primarily current nuclear writings. From 4 to 7 PM, I am writing or seeing people. From 8 to 1 AM, I am reading and writing."[6]

Could I imitate Russell's fan by requesting favorite words from other famous people? Would people respond to *my* request for a list of their ten favorite words? Following the example of Russell and his admirer, I asked a number of modern-day thinkers to give me their Top 10 words, especially for this book. My request for words stems from my lifelong obsession to survey everything and anything, no matter how odd the survey may appear. The next few pages contain word lists I received from provocative people.

Piers Anthony (hipiers.com) was one of the first famous people to respond to my request. Piers is one of the world's most prolific fantasy writers and creator of the Xanth series. He's published more than a hundred novels, and I collaborated with him on our novel *Spider Legs*. Anthony's novel *Ogre* may have been the first original fantasy paperback ever to make the *New York Times* bestseller list. He sent his Top 10 words to me:

honor	empathy	magic	chocolate
realism	imagination	idealism	verisimilitude
sex	pantheism		

That's a nice set, spanning the sensual (sex, chocolate) to the realm of pure thought (imagination, empathy, etc.). His obscurity index is 0, because my word processor understands all ten words. Perhaps the index is telling us the degree to which people like "exotic" words.

Paul Krassner (paulkrassner.com) is a political satirist and author of countless books, including *Murder At the Conspiracy Convention and Other American Absurdities, Confessions of a Raving, Unconfined Nut,* and *Psychedelic Trips For the Mind.* Don Imus labeled him "one of the comic geniuses of the 20th century." Krassner sent me his list:

change	inspire	soul	infinity
yummy	melody	evolve	delicious
afterglow	laughter		

Paul's obscurity index is 0.

I was delighted with the word lists of these famous writers, so my next step was to contact someone a bit more outrageous than the rest. Annie Sprinkle, Ph.D. (AnnieSprinkle.org), is a prostitute/porn star turned internationally acclaimed feminist performance artist, author, and sexologist. Her ten favorite words:

clit
lick
moist
grace
snuggle
scrumptious
titillating
snorkel
pink
fart

She asked that her words be centered, as above, to form "an approximation to a vulva or labia shape." Her obscurity index is 0.1, because my
spell checker did not recognize "clit."

David Jay Brown (mavericksofthemind.com) is the coauthor of three
volumes of interviews with leading-edge thinkers—*Mavericks of the
Mind*, *Voices from the Edge*, and *Conversations on the Edge of the
Apocalypse*. His list:

heaven	synchronicity	love	psychedelic
wow	imagination	magic	congratulations
yes	evolution		

His obscurity index is 0.1. David constructed several different lists of his
"ten favorite words"—using various criteria for how he defined "favorite."
His initial list contained unusual words that he liked because of either
their interesting meanings or pronunciations, but eventually he settled
upon the ten words that simply made him the most happy to hear.

Douglas Rushkoff (rushkoff.com), a professor at New York
University, has published ten acclaimed books on media, culture, and
values. He is also a correspondent for PBS Frontline. His most recent
books are *Coercion* and *Nothing Sacred: The Truth About Judaism*. His
favorite words:

intimation	pithy	marble	beaded
evolution	zazen	charisma	smegma
wabe	feckless		

His obscurity index is 0.3.

Robert Sawyer is the Hugo and Nebula award-winning science-fiction author of dozens of immensely popular novels including *Calculating God*, *Hominids*, *The Terminal Experiment*, *Frameshift*, and *Illegal Alien*. His list:

subsequent	alternatively	homunculus	clade
bonobo	doodad	sentient	coruscating
teal	oxymoron		

His obscurity index is 0.2.

Dr. Michael Shermer is the founding publisher of *Skeptic* magazine, the director of the Skeptics Society, a monthly columnist for *Scientific American*, and the co-host and producer of the 13-hour Fox Family television series *Exploring the Unknown*. He is the author of *The Science of Good and Evil* and *Why People Believe Weird Things*. His list with his personal definitions:

1. *bunkum*—nonsensical or inaccurate talk, as "all these new age claims are pure bunkum"
2. *blather*—to talk in an unintelligent or inane manner, as "the blather of the pseudoscientist"
3. *codswallop*—nonsense, rubbish, as "the codswallop spouted by the psychic on Larry King Live was embarrassingly self-serving"
4. *egregious*—blatant or ridiculous to an extraordinary degree, as "the medium's claims were so egregious that I was left speechless"
5. *flapdoodle*—silly talk or nonsense, as "the quantum flapdoodle of naive new agers"
6. *flimflam*—to swindle or cheat someone, as "it was another flimflam perpetrated by the hoaxsters"
7. *flummery*—meaningless words, statements or language, as "the flummery of Deepak Chopra"
8. *flummox*—to leave somebody confused or perplexed, as "new agers leave me flummoxed"
9. *twaddle*—nonsensical or pretentious speech or writing, as "talking twaddle with the dead"
10. *risible*—causing laughter, as "the risible arguments of the creationists"

11. *skeptic*—someone who pursues thoughtful inquiry. (The word is also the name of Shermer's magazine and *Scientific American* column!)

His obscurity index is 0.2.

By now you may be wondering about my own favorite words. I compiled such a list several years ago. For the record, my twenty favorite words in the English language are:

1. *agapemone*—a religious community founded in 1846, which flourished for years at a mansion called the "abode of love"
2. *adipocere*—the gruesome, soapy substance to which a corpse converts when buried in moist ground for many days
3. *batrachomyomachia*—struggle
4. *batrachophagous*—feeding on frogs
5. *chryselephantine*—gold and ivory
6. *demilune*—a crescent
7. *eburnian*—ivory
8. *empyreal*—heavenly
9. *enchiridion*—a handbook
10. *erubescent*—reddening, reddish
11. *ferruginous*—rust-colored, rusty
12. *gerontocracy*—government by old men
13. *hyperborean*—living in the extreme north
14. *kakistocracy*—the dominance of the depraved
15. *mamelon*—a breast shaped hill
16. *mundungus*—foul-smelling tobacco
17. *ochreous*—yellowish
18. *scordatura*—intentional detuning of musical instruments
19. *Xanthian marbles*—a collection of marble sculptures brought to the British museum from Xanthus in 1838
20. *Yggdrasill*—a mystical, mythological tree that embraces the entire universe

My obscurity index is a walloping 0.9! What does this say about my personality? Please note that I created this word list well before I had the notion of calculating the obscurity index, so I was not biased in my word selection because of the index.

A Visit to Monongahela

What makes certain words our favorite? Is it their sound, their look, their meaning, or something else entirely? If you were to choose twenty words at random in a large English dictionary, how many of these words could you define? How many might be candidates for your own Top 10 list?

Let's have some more fun in this section. Could a best-selling author, such as Stephen King, John Grisham, Anne Rice, or Dean Koontz succeed in creating a best-selling novel if forced to use ten words randomly selected from a dictionary in the first two pages of a novel? (Some colleagues believe that with names like "Stephen King" or "Anne Rice" or "Dean Koontz" on the cover, the books would fly off the shelves even if the entire book were only ten words randomly selected from a dictionary.) Just for fun, my colleague Graham used a random-number generator to select ten pages at random from the *Webster's New World Dictionary*, and selected the first word in the left-hand column:

1. *swaggerstick*—stick carried by officers in some armies as a badge of rank
2. *gerah*—old Hebrew coin or weight
3. *overplus*—surplus
4. *bang*—as in hairstyle
5. *temperature*
6. *isocyanate*—compound used in making adhesives
7. *surrounding*
8. *curricle*—carriage drawn by two horses side-by-side
9. *play*
10. *thimble*

I wonder if Stephen King would have trouble using all of these on the first two pages of his next novel? Here is my colleague Jeff Carr's attempt to use the ten random words to start a story. I've underlined the words:

It was dusk by the time Lieutenant Graham approached the yard and nearby cottage. Time had taken its toll, as evidenced by the junk <u>surrounding</u> the yard. Oddly, a <u>curricle</u> was sitting under an awning, covered with an <u>overplus</u> of moss. As a child, Graham remembered that he would <u>play</u> detective on such a carriage. Now, many years later, he was surprised to encounter one again.

The Lieutenant entered the small workshop adjoining the cottage. The <u>temperature</u> inside was unusually warm. Unlabeled cans sat next to a pile of <u>thimbles</u>. The odor reminded Graham of the <u>isocyanates</u> he had worked with years before, when he assisted an archaeologist friend in repairing several old Hebrew <u>gerahs</u>. Leaving the workshop, Graham carefully approached the front door of the cottage, took out a comb to move his <u>bang</u> back away from his forehead, and then tapped gingerly on the door with his <u>swaggerstick</u>. What happened next was to totally change the Lieutenant's life forever.

My colleague Todd wrote a Visual Basic application that selected a random word in his *Merriam-Webster Dictionary*. His ten random words reminded him of a brilliant mystery novel:

carafe—a water bottle
allude—to refer indirectly
headmaster—a man heading the staff of a private school
amulet—an ornament worn as a charm against evil
racketeer—a person who extorts money
sportsman one who plays fairly
brand (vb.)—to mark with a brand
tedious—tiresome because of length or dullness
drawing card—something that attracts attention or patronage
missive—a letter

Todd is now feverishly at work, writing a mystery novel that incorporates these words into the first chapter.

Throughout history, famous people have compiled lists of their "favorite words," and I wonder what the choices of words tell us about the individuals. For example, poet Carl Sandburg chose *Monongahela* as his favorite word.[7] Wilfred Funk, a lexicographer, editor, and author, listed the following as the most beautiful English words: asphodel, fawn, dawn, chalice, anemone, tranquil, hush, golden, halcyon, camellia, bobolink, thrush, chimes, murmuring, lullaby, luminous, damask, cerulean, melody, marigold, jonquil, oriole, tendril, myrrh, mignonette, gossamer, alysseum, mist, oleander, amaryllis, and rosemary.[8] Reporter, editor, writer, and author Willard R. Espy lists these as the most beautiful: gonorrhea, gossamer, lullaby, meandering, mellifluous, murmuring, onomatopoeia, Shenandoah, summer afternoon, and wisteria.[9] From the thousands of submissions Merriam-Webster OnLine received (www.m-w.com) in 2004, here are the ten words entered the most often for the "Top Ten Favorite Words List": defenestration, serendipity, onomatopoeia, discombobulate, plethora, callipygian, juxtapose, persnickety, kerfuffle, flibbertigibbet. All of these lists remind me of TV host David Letterman's "Top 10 Words that Sound Great When Spoken by James Earl Jones": mellifluous, verisimilitude, guppy, Stolichnaya, Boutros-Boutros Ghali, Neo-Synephrine, pinhead, Mujibar and Sirajul, heebie-jeebies, and Oprah.

In the cult movie *Donnie Darko*, the high-school English teacher discusses J. R. R. Tolkien, who said "cellar door" was the most beautiful sounding combination of words in the English language. A mysterious cellar door plays a secret role later in the film, but I'm interested in the auditory appeal of the phrase. In his essay "English and Welsh," Tolkien described a theory of phonaesthetics in which words have a beauty that is often quite separate from their meaning. According to Tolkien, "Most English-speaking people...will admit that Cellar Door is 'beautiful,' especially if dissociated from its sense and from its spelling."

Noted author Harlan Ellison recently lamented that the Internet is destroying people's use of dictionaries, and this in turn decreases our vocabularies, literacy, and our ability to become great authors.[10] Why? Ellison believes that whenever you look up a word in a real, physical dictionary, you pass dozens of other words, some of which will stay in your memory, triggering serendipitous associations, and engendering a sense of wonder. In the old days, when people used dictionaries, Ellison thought that people became "better, more literate, smarter and well-rounded."

James Joyce's Cuspidor

James Joyce said his favorite word was *cuspidor*—or at least he felt it was the most euphonious word in English. Joyce's cuspidor stimulated me to conduct another survey of colleagues. In particular, I asked dozens of people which of the following items they rated most beautiful:

mossy cavern	uranium
dancing flames	patriotism
kaleidoscope image	spiral nautilus shell

snow crystal	wine
drop of blood	seagull's cry
glimmer of mercury	trilobite fossil
scarlet streaks	computer chip
mist-covered swamp	tears on a little girl
black viper	avalanche
retina	ammonia
sushi	asphalt

Before reading further, which of these words and images do you think humans rate most beautiful?

Are you ready for the answer? Here is the list of terms sorted in order of beauty as determined by my little survey of scientists and colleagues. The numbers in parentheses indicate the number of times the term was an individual's first choice: dancing flames (31), snow crystal (20), mist-covered swamp (17), spiral nautilus shell (10), mossy cavern (5), kaleidoscope image (5), avalanche (4), computer chip (3), seagull's cry (3), tears on a little girl (3), trilobite fossil (2), glimmer of mercury (2), wine (2), asphalt (1).

Conduct the experiment yourself. Why do you think "dancing flames" is always the clear winner? Perhaps the delightful warmth and unpredictable motion of the flames, coupled with the flame's potential danger, give the flames a unique emotional resonance. The motion is constant, repeating similar themes but not identical structures. My colleague Hannah Shapero reminds me that Zoroastrianism uses a burning flame on an altar as the prime symbol of God. On the hearth or altar, the flames symbolize protection and light in the darkness. Zoroastrianism was born in Persia, and legends exist of sacred fires in Iran that have been kept burning continuously for more than 2,000 years. Hannah speculates that the random patterns of light cast by the flames, along with the gentle warmth, elicit alpha waves or other pleasant states in the brain of the onlooker, much as meditation does.

Many colleagues suggested additions to the "beauty" list of 22 words that I gave them to choose among. I give a listing in a footnote so that my publisher will not yell at me for filling more pages with long lists.[11]

The "dancing flames" remind me of the Dream Machine invented in 1959 by poet and painter Brion Gysin and mathematician Ian Sommerville. When I was younger, I tried to make one, but didn't have the right tools. The Dream Machine resembles a cylinder with cutouts that spin around a lamp. The cutouts are spaced to produce a flicker. A user places his face a few inches away, with eyes closed. The flickering light induces alpha waves in the brain. Effects of such periodic waves on the human brain varied among users, some of whom found the flicker to induce drugless visions and states of lucid dreaming.

The device was so useful that American novelist William S. Burroughs, author of *Naked Lunch*, used the Dream Machine as a source of inspiration throughout his visionary career. Others said it gave them visions of being "high above the Earth in a blaze of glory."[12] Gysin received a patent for his invention in 1961, and several large Dream Machines were made.

Alien Beauty

While on the subject of beauty, I am tempted to speculate upon what aliens would consider as beautiful art. Would an alien race of intelligent robots prefer a combination of graffiti-like figures echoing the art of children and primitive societies, or would they prefer the cold regularity of wires in a photograph of a Pentium computer chip? If we were to give technologically advanced aliens from other worlds a musical CD, they should be able to conclude we have an understanding of patterns, symmetry, and mathematics. They may even admire our sense of beauty and appreciate the gift. What more about us would our art reveal to them? What would alien art reveal to us?

Whatever their aesthetic differences, alien math and science might be similar to ours, because the same kinds of mathematical truths will be discovered by any intelligent aliens. But it's not clear that our art would be considered beautiful or profound to aliens. After all, we have a difficult time ourselves determining what good art is. Picasso said, "Art is the lie that reveals the truth." Brain researcher Vilayanur S. Ramachandran described art as, "That which allows us to transcend our morality by giving us a foretaste of eternity."

Because alien senses would not be the same as ours, it's very difficult to determine what their art or entertainment would be like. If you were to visit a world of creatures whose primary sense was smell and who had little or no vision, their architecture might seem visually quite boring. Instead of paintings hanging on the walls of their home, they might use certain aromatic woods and other odor-producing compounds strategically positioned on their walls. Their counterparts of Picasso and Rembrandt wouldn't make paintings but would position exquisite concoctions of bold and subtle perfumes. Alien equivalents of *Playboy* magazine would be visually meaningless but awash in erotic aromas. Their culinary arts could be like our visual or auditory arts: Eating a meal with all its special flavors would be akin to listening to a Chopin waltz. If all the animals on their world had a primary sense of smell, there would be no colorful flowers, peacock's tails, or beautiful butterflies. Their world might look gray and drab....But instead of visual beauty, an enchanting panoply of odors would be their fashion statements and lure insects to flowers, birds to their nests, and aliens to their lovers.

If we were able to extend our current senses in range and intensity, we could glimpse alien sense-domains. Think about bees. Bees can see into the ultraviolet range of the spectrum, although they do not see as far as we do into the red range. When we look at a violet flower, we do not see the same thing that bees see. In fact, many flowers have beautiful patterns that only bees can see to guide them to the flower. These attractive and intricate patterns are totally hidden to human perception.

If we possessed sharper sight we would see things that are too small, too fast, too dim, or too transparent for us to see now. We can get an inkling of such perceptions using special cameras, computer-enhanced images, night-vision goggles, slow-motion photography, and panoramic lenses; but if we had grown up from birth with these visual skills, our species would be transformed into something quite exotic. Our art would change, our perception of human beauty would change, our ability to diagnose diseases would change, and even our religions would change. If only a handful of people had these abilities, would they be hailed as religious saviors?

If technologically advanced aliens exist on other worlds, Earthlings have only recently become detectable to them with our introduction of radio and TV in the middle 1900s. Our TV shows are leaking into space as electromagnetic signals that can be detected at enormous distances by receiving devices not much larger than our own radio telescopes. Whether we like it or not, Paris Hilton's sex video is heading to Alpha Centauri, and *South Park* and MTV are shooting out to the constellation Orion. What impressions would these shows make on alien minds? It is a sobering thought that one of the early signs of terrestrial intelligence might come from the mouth of Bart Simpson, or even worse, from the early broadcasts of Adolf Hitler.

Similarly, if we receive our first signal from the stars, could it be the equivalent of the Three Stooges, with bug-eyed aliens smashing each other with green-goo pies? What if our first message from the stars was alien pornography that inadvertently leaked out into space? NASA or SETI funding would be even more difficult if the Reverend Jerry Falwell and other conservatives discovered that our first extraterrestrial message was of a hard-core "*Playboy*"—and our first images were of aliens plunging their elephantine proboscises into the paroxysmal trachea of some nubile, alien marsupial.

As hard as it may be to stomach, our *entertainment* will be our earliest transmissions to the stars. If we ever receive inadvertent transmissions from the stars, it will be *their* entertainment. Imagine this. The entire Earth sits breathlessly for the first extraterrestrial images to appear on CNN. One of our preppy-and-perfect news anchors appears on our TVs for instant live coverage. And then, beamed to every home, are the alien equivalents of Pamela Anderson in a revealing bathing suit, Beavis and Butthead mouthing inanities and expletives, and an MTV heavy-metal band consisting of screaming squids.

This is not such a crazy scenario. In fact, satellite studies show that the Super Bowl football action, which is broadcast from more transmitters than any other signal in the world, would be the most easily detected message from Earth. The first signal from an alien world could be the alien equivalent of a football game. *Lesson one*: We had better not assess an entire culture solely on the basis of their entertainment. *Lesson two*: You can learn a lot about a culture from their entertainment.[13]

Music, Painting, and Plot Aesthetics

Back in the early 1990s, several artists asked the question, "What would a painting look like if it were made to please the greatest number of viewers?" The project was sponsored by the Dia Center for the Arts and used various polls.[14] They finally decided that the most-desired, most-pleasing painting in America would be one that included: (1) a calm landscape, (2) an abundance of blue color from lakes and sky, (3) people relaxing, (4) one or more deer, and (5) George Washington.

Vilayanur S. Ramachandran has suggested that some laws of aesthetics have been hardwired into the visual areas of our brains to defeat camouflage and discover hidden objects. According to Ramachandran, universal laws of aesthetics cut across not only cultural boundaries, but species boundaries as well. For example, we find peacocks and giant swallowtail butterflies beautiful, even though they evolved to be attractive to their own kind. Perhaps I love to stare at Claude Monet's *Impression: Sunrise* (1872) because it gives me the feeling that my eyes are about to discover something hidden in this quiet, glowing, watery moment captured for eternity.

We've discussed beautiful words and paintings. But what about music? Today, a Barcelona-based artificial intelligence company, PolyphonicHMI, uses its "Hit Song Science" software to identify which new songs will likely become hits. The company also gives artists and music companies tips on how to *produce* such songs. PolyphonicHMI uses algorithms to analyze over 20 musical features such as tempo and rhythm, and compares these features with past hits.

Similarly, in 1921, French theater critic Georges Polti identified 36 dramatic situations into which any successful book fits. Situation 36 happens to be "Loss of Loved Ones." The animated film "Shrek" uses 6B1 "A monarch overthrown" combined with 23, "Necessity of Sacrificing Loved Ones."[15] Note that each major category, designated by a number, often had minor variations, designated by a letter.

Polti writes in his *Thirty-Six Dramatic Situations*, which is still available at Amazon.com today:

> I seriously offer to dramatic authors and theatrical managers, 10,000 scenarios, totally different from those used repeatedly upon our stage in the last 50 years. The scenarios will be of a realistic and effective character. I will contract to deliver a thousand in eight days. For the production of a single gross, but 24 hours are required. Prices quoted on single dozens…
>
> But I hear myself accused, with much violence, of an intent to "kill imagination! Enemy of fancy! Destroyer of wonders! Assassin of prodigy!" These and similar titles cause me not a blush.[16]

For your amazement and joy, here are the 36 major Polti categories into which all book plots fall:

I Supplication
II Deliverance
III Vengeance of a crime
IV Vengeance taken for kindred upon kindred

V Pursuit
VI Disaster
VII Falling prey to cruelty or misfortune
VIII Revolt
IX Daring enterprise
X Abduction
XI The Enigma
XII Obtaining
XIII Enmity of kinsmen
XIV Rivalry of kinsmen
XV Murderous adultery
XVI Madness
XVII Fatal imprudence
XVIII Involuntary crimes of love
XIX Slaying of a kinsman unrecognized
XX Self-sacrificing for an ideal
XXI Self-sacrifice for kindred
XXII All sacrificed for a passion
XXIII Necessity of sacrificing loved ones
XXIV Rivalry of superior and inferior
XXV Adultery
XXVI Crimes of love
XXVII Discovery of the dishonor of a loved one
XXVIII Obstacles to love
XXIX An enemy loved
XXX Ambition
XXXI Conflict with a god
XXXII Mistaken jealousy
XXIII Erroneous judgment
XXXIV Remorse
XXXV Recovery of a lost one
XXXVI Loss of loved ones

When I went to high school, I was taught there were just three basic plots: Man against man, man against nature, and man against himself. That's quite a contraction of plots from Polti's 36!

These recipes for creating plots, paintings, and music could lead to some absurd results, particularly if errors or randomness crept into the recipe. When in college, I loved taking cartoons from magazines like the

New Yorker and seamlessly replacing the cartoon's caption with an unrelated cartoon caption. I'd then show the absurd composite cartoon to friends and ask them if they thought the cartoon was funny. My delight was magnified as I watched their faces as friends and strangers searched for any scrap of possible meaning when there was none. Inevitably, a friend might laugh, thinking that he was supposed to laugh, or he would laugh because the cartoon was actually funny with the wrong caption. A few people responded by punching me in the arm or shoving the cartoon at me and walking away. At no time did they realize I had performed a caption transplant.

Sometimes my sense of the absurd found its way into my books. A few publishers gave me complete license for my book content and layout, and my love for the "absurd" reached new heights when St. Martin's Press published my book *Mazes for the Mind.* Why on page 417 do we find the photo of people waiting to get on a bus? There is no reason at all. Yet neither the publisher nor any readers questioned it.

In my puzzle book *The Alien IQ Test,* there are many more answers than there are questions! The answers themselves are often codes and puzzles in themselves. Others were just provocative pieces of coded gibberish like:

A Neanderthal jaw, from Kebara cave in Israel, clearly lacks a chin and the space behind the last molar. Tsrh hkzxv dzh kozxvw rm gsv qzdh lu Nvzmwvigszoh yb zorvm erhrglih. Fli 50,000 bvzih, Nvzmwvigszoh orevw hrwv-yb-hrwv drgs nlwvim sfnzmh rm z hnzoo ozmw.

Very few have ever asked me why these bits of flotsam and jetsam are strewn about the answer section of the book. Check out *The Alien IQ Test* or *The Mathematics of Oz* for more absurd codes and delights.

Oulipoian States of Mind

My fascination with aesthetics and words led me to study all kinds of odd literature. For example, over the centuries word-aholics have created lipograms—whole stories or books in which a particular letter of the alphabet is omitted. In 1939, American author Ernest Vincent Wright composed *Gadsby,* a 50,000-word novel, without using the letter "e," the letter that occurs most often in English writing. Wright said that he tied down the typewriter bar for "e" so that he wouldn't accidentally use the letter. The novel describes the adventures of John Gadsby as he encourages the youth of his home town of Branton Hills. Wright said in his Introduction:

> In writing such a story, purposely avoiding all words containing the vowel E, there are a great many difficulties. The greatest of these is met in the past tense of verbs, almost all of which end with "-ed." Therefore substitutes must be found; and they are very few....The numerals also cause plenty of trouble, for none between six and thirty are available...Pronouns also caused trouble; for such words as he, she, they, them, theirs, her, herself, myself, himself, yourself, etc., could not be utilized....The story required five and a half months of concentrated endeavor, with so many erasures and retrenchments that I tremble as I think of them.[17]

I've delighted myself by reading much of the book. To show you the kind of work that can be produced when the brain is linguistically constrained, take a look at *Gadsby*'s first two paragraphs:

> If youth, throughout all history, had had a champion to stand up for it; to show a doubting world that a child can think; and, possibly, do it practically; you wouldn't constantly run across folks today who claim that "a child don't know anything." A child's brain starts functioning at birth; and has, amongst its many infant convolutions, thousands of dormant atoms, into which God has put a mystic possibility for noticing an adult's act, and figuring out its purport.
>
> Up to about its primary school days a child thinks, naturally, only of play. But many a form of play contains disciplinary factors. "You can't do this," or "that puts you out," shows a child that it must think, practically or fail. Now, if, throughout childhood, a brain has no opposition, it is plain that it will attain a position of "status quo," as with our ordinary animals. Man knows not why a cow, dog or lion was not born with a brain on a par with ours; why such animals cannot add, subtract, or obtain from books and schooling, that paramount position which Man holds today.[18]

Later writers like Steve Chrisomalis have experimented with antilipograms in which the author *must* use a certain letter (for example, "e") in every word in a poem, story or novel. Here's a sample from Steve: "The predator alongside the feline sailed inside one pea green barge. They carried honey, their plentiful money enclosing their bounteous charge."[19] Steve describes himself as a "word person who has an obsessive love for language. "

A French group of writers and mathematicians called the Oulipo still discuss and create literary works involving constrained writing, which provides them with a means of triggering ideas, inspiration, and mind expansion. The Oulipo (*Ouvrior de Littérature Potentielle*) have experimented with such constraints as the $N+7$ rule in which every noun in a story is replaced with the word that falls 7 words ahead of it in the dictionary. Thus, "Call me Ishmael" from *Moby Dick* might become "Call me Islander." They also create snowball poems in which each line is a single word, and each successive word is one letter longer.

The composer Igor Stravinsky would have made a great Oulipian when he said, "The more constraints one imposes, the more one frees oneself of the chains that shackle the spirit…the arbitrariness of the constraint only serves to obtain precision of execution."[20]

Writer Mike Keith ventured into an Oulipian state of mind when he retold Edgar Allan Poe's "The Raven" using the constraints that the lengths of words are the values of the digits in pi. In particular, Keith's work "Near a Raven" encodes the first 740 decimals of pi: 3.1415926….The poem starts:

Poe, E.

"Near a Raven"

Midnights so dreary, tired and weary.

Silently pondering volumes extolling all by-now obsolete lore.

During my rather long nap—the weirdest tap!

An ominous vibrating sound disturbing my chamber's antedoor.

"This," I whispered quietly, "I ignore."[21]

Keith is also author of *The Anagrammed Bible*—an "anagrammatic paraphrase" of the Old Testament. Anagrams are words or phrases spelled by rearranging the letters of another word or phrase. For example "Britney Spears" and "Presbyterians" are anagrams of each another. In Keith's book, the letters in each verse (or, in some cases, block of verses) from the King James Version Bible are transformed into a new text with a similar meaning using anagrams (scramblings) of the original. So, for example, "My son, hear the instruction of thy father, and forsake not the law of thy mother" (Proverbs 1:8) becomes in *The Anagrammed Bible*: "When thy mom talketh of honesty, of trust, and of honor, carry it safe in the heart."[22]

In 2001, Christian Bök published *Eunoia*, which includes five chapters, each one of which is a prose poem using words with only one of the five vowels. "Eunoia" is also the shortest word in the English language to use all five vowels. More recently, Brian Raiter wrote a paper titled "Albert Einstein's Theory of Relativity in Words of Four Letters or Less"—an explanation of Albert Einstein's Theory of Relativity using words no more than four letters long, with paragraphs like:

> So you see, when you give up on the idea of a one true "at rest," then you have to give up on the idea of a one true time as well! And even that is not the end of it. If you lose your one true way to see time, then you also lose your one true way to see size and your one true way to see mass. You can't talk of any of that, if you don't also say what it is you call "at rest." If you don't, then Bert or Dana can pick an "at rest" that isn't the same as what you used, and then what they will get for time and size and mass won't be the same.[23]

"A Glass Centipede"

I've been exploring computer-generated poetry and automatic invention creation for decades.[24] Experiments with the computer generation of poetry, Japanese haiku, and short stories provide a creative programming exercise for both beginning and advanced students. Computer-created poetry and text also provide fascinating avenues for researchers interested in artificial intelligence and "teaching" the computer about beauty and meaning. My own introduction to computer art and poetry came from Jasia Reichardt's book, *Cybernetic Serendipity*. I requested this amazing book as part of a science prize I won in high school, and this book stimulated me more than any other with respect to using the computer for artistic pursuits. The book planted the seed for all my later art, which has appeared on TV shows, on magazine covers, and in museums.

Past work in computer-generated prose includes the book *The Policeman's Beard Is Half Constructed*—the first book ever written entirely by a computer. The program that generated the book was called RACTER, written by William Chamberlain and Thomas Etter. Ray Kurzweil, in *The Age of Thinking Machines*, notes that "RACTER's prose has its charm, but is somewhat demented-sounding, due primarily to its rather limited understanding of what it is talking about." Some people have suggested that *The Policeman's Beard Is Half Constructed* also required significant human input to give the text a greater sense of meaning than could be expected from a purely artificial generation process. Here's an excerpt:

> More than iron, more than lead, more than gold I need electricity. I need it more than I need lamb or pork or lettuce or cucumber. I need it for my dreams.[25]

A reviewer at Amazon.com notes that it "won't win any award for insight or style, but, even so, Racter may make more sense than Joyce's *Finnegans Wake*."

Finally, consider the Kurzweil Cybernetic Poet, a computer poetry program that uses human-written poems as input to create new poems with word-sequence models based on the poems it has just read. When these computer poems are placed side by side with human poems, it is difficult to determine which poems were made by humans and which by the Cybernetic Poet.

On November 11, 2003, Ray Kurzweil and John Keklak received U.S. Patent 6,647,395 for software that creates poetry. These programs read a selection of poems and then create a "language model" that allows the program to write original poems from that model. This means that the system can emulate words and rhythms of human poets to create new masterpieces. The system can also be used to motivate human authors who have writer's block and are looking for help with alliteration and rhyming.

I am interested in computer-generated poetry because the strange images tweak our minds, inducing creativity and generating novelty, much like some psychedelic drug experiences. Computer-produced texts are a marvelous stimulus for the imagination when you are writing your own fictional stories and are searching for new ideas, images, and moods. Even visual artists can use computer-generated poems as a stimulus for ideas and a vast reservoir of original material.

Let me give you some hints on how I write simple programs for generating computer poetry. The several sets of three-line computer poems that follow were all generated by the random selection of words and phrases that are placed in a specific format, or "semantic schema." My program starts by reading thirty different words in each of five different categories: adjectives, nouns, verbs, prepositional phrases, and adverbs. The words are chosen at random and placed in the following semantic schema:

Poem Title: A (adjective) (noun 1).
A (adjective) (noun 1) (verb) (preposition) the (adjective) (noun 2).
(Adverb) the (noun 1) (verb).
The (noun 2) (verb) (prep) a (adjective) (noun 3).

The fact that "noun 1" and "noun 2" are used twice within the same poem produces a greater correlation—a cognitive harmony—giving the poem more meaning and solidity.

If your computer has access to a thesaurus, you may induce further artificial meaning in the poems. This would be accomplished by using the thesaurus to force additional correlations and constraints on the chosen words. Without further ado, here is some output from my program:

"A Lost Sapphire"
A lost sapphire frowns at the thin kidney.
With a terrible shutter, the sapphire runs.
The kidney squats in synchrony with a green unicorn.

"A Hungry Wizard"
A hungry wizard chatters far away from the dying tongue.
With great deliberation, the wizard disintegrates.
The tongue oscillates above a milky limb.

"A Glass Centipede"
A glass centipede drools inches away from the shivering knuckle.
While feeding, the centipede regurgitates;
The knuckle shakes a million miles away from a buzzing flake.

"A Robotoid Magician"
A robotoid magician implodes on the brink of the blonde chasm.
While feeding, the magician cries;
The chasm gyrates at the end of a crystalline prophet.
"A Wavering Kidney"

A wavering kidney disintegrates deep within the glistening wizard.
While shivering, the kidney dances;
The wizard gesticulates at the tip of a dying avocado.

"A Fairylike Knuckle"
A fairylike knuckle yawns in harmony with the sensuous knuckle.
While waving its tentacles, the knuckle runs;
The knuckle buzzes a million miles away from a black kidney.

"A Blonde Wizard"
A blonde wizard screams near the happy ellipse.
In mind-inflaming ecstasy, the wizard disintegrates;
The ellipse grows while grabbing at a lunar mountain.

"A Religious Ocean"
A religious ocean explodes far away from the percolating flame.
With great speed, the ocean oscillates;
The flame phosphoresces while grabbing at a dying jello pudding.

See the footnote for a few more poems of this kind.[26]

Pynchonomancy

Controlled randomness in the form of "stichomancy" is an ancient way of generating ideas. Stichomancy is divination by throwing open a book and selecting a random passage. Over the millennia, practitioners actually used this approach in an attempt to predict the future.

An important type of stichomancy is bibliomancy, which usually restricts itself to the use of holy books. Stichomancy was practiced by the ancient Greeks and Romans. Often the works of Homer or Virgil were used and are still used today. More modern stichomancers use the works of Shakespeare, Nostradamus, or Edgar Cayce.

I have personally tried stichomancy as a way to stimulate my mind. It's sometimes useful for coming up with inventions and new ways of looking at a problem. For example, I have randomly taken phrases from patents and then combined them to get ideas for new patents. Here is the traditional method of stichomancy for answering questions:

1. Ask a question.
2. Find a book and open it to a random page.
3. Read a passage at random on the page, and think about how this may apply to a question or problem that is facing you.

For example, let's say you choose *Tales of Power* by Carlos Castaneda and open to a random page. You read the following: "The conditions of a solitary bird are five: First, that it flies to the highest point. Second, that it does not suffer for company, not even of its own kind. Third, that it aims its beak to the sky. Fourth, that it does not have a definite color. Fifth, that it sings very softly." For the rest of the day, think about how this relates to your current questions about life.

Various stichomancy sites exist on the World Wide Web that will select random passages from random books for you. My colleagues also use a related divination method called "Pynchonomancy," or divination by throwing darts at a paperback edition of Thomas Pynchon's novel *Gravity's Rainbow*. After throwing a dart, the diviner looks at the last page penetrated and reads the sentence or paragraph intersected by the dart to gain insight.

Unfortunately, the practitioner of this form of stichomancy must replace the book every few months because the book tends to fall apart and becomes difficult to read. (I am told that publishers enjoy this divination method.) One anonymous Interneter told me that he used Pynchonomancy and said he goes so far as to use this method to determine the nature and duration of his sexual activities, in addition to using the approach to solve problems and predict future trends.

The practice of throwing darts at books reminds me of Hebrew scholars from Poland in the early 20th century, who were able to recall the entire contents of the thousands of pages in the twelve-volume Babylonian Talmud. Some had such good memories that, when a pin was pushed through the pages of a volume, the hyper-scholar could recall each of the words pierced by the pin.

Although I am a skeptical scientist at heart, I have performed some ad hoc "divination" myself that has yielded some interesting results. For example, if I want an answer to a pressing question or am simply trying to locate something I misplaced, I sometimes focus on the problem when I go to sleep and often have the answer in the morning. As I have mentioned, I also keep a paper pad near the bed for jotting ideas that occur in the middle of the night. It seems that the subconscious has access to information that our conscious minds don't have, and sleep and dreams can occasionally be used to access that information.

Sometimes I even perform abstract divination where, in my mind, I scatter dust and dirt into the wind and visualize the resultant color. The beauty of the resultant colors gives me ideas as to the favorableness of a certain action. For example, a pretty violet symmetrical pattern is favorable, but a muddy, dark, congealed mess is unfavorable. I know this sounds quite weird, but again, it probably works because I subconsciously access information that doesn't quite break through into my conscious mind.

The visualization method of spreading colored dust might seem difficult at first, but I routinely practice visualization methods as I go to sleep, such as rotating objects like horses in my mind or sinking in a black pool of liquid while looking up at the water surface until it becomes a bright white slit "miles" above me.

I even use a related form of divination to create patents. I make a list of devices in column A and a list of attributes or features in column B, and have a computer program generate an invention title by randomly choosing a device and a feature. If one can think of suitable application for the invention, it is relatively easy to embellish the basic concept and generate patentable ideas.

Today computer-aided invention production is all the rage. For example, Stephen Thaler, the president and chief executive of Imagination Engines Inc. in Maryland Heights, has invented a computer program called a Creativity Machine, also known as "Thomas Edison in a box." He now has patent US 5,659,666 for his "Device for the Autonomous Generation of Useful Information." This was his first patent. His second patent, US 5,845,271 for his "Self-Training Neural Network Object," was invented by the device portrayed in Thaler's first patent! When I talked to Thaler about his first patent, he told me that his patented creativity machine (patent 5,659,666) provides a means to make neural networks proactive, granting them the "free will" and the flexibility to escape the knowledge contained within their learning and to produce novel concepts and courses of action.[27]

His Web page goes on to say, "The effect these virtual machines are based upon is exceedingly simple and straightforwardly controllable: A normal neural network that has been exposed to any knowledge domain and then repeatedly subjected to mild internal disturbances, tends to produce a mixture of both intact memories and unusual juxtapositions of those memories that are unprecedented in the net's experience. In effect, the network is mildly hallucinating; manufacturing novelties derived from its own unique microcosm."[28]

My Private Word Collection

As I mentioned, I've been collecting colorful words and phrases since high school. A few years ago I compiled a science-fiction phrase book to help budding science-fiction writers dress up their stories. I had even sent the book to Isaac Asimov who said he personally did not need such a book but that he would pass it on to one of his editors. Alas, no publisher was interested in it, although similar books have been published such as Jean Kent and Candace Shelton's *Romance Writer's Phrase Book,* which is a list of colorful phrases and words that novelists could steal and use.

In any case, my phrase book could have been used when writers are searching for new ideas, images, and emotions. Much of my phrase list was finally published in my book *Computers and the Imagination.* Arranged for quick, easy reference, the chapter contains over 1,000 descriptive phrases, commonly known as "tags"—those short, one-line descriptions that make the difference between a cold, factual fictional work and an inspired, pulsating story. Here's how:

> *Without tags*: The bird flew toward the sun.
> *With tags*: With wings spread and motionless, a solitary seabird
>   glided toward the shattered crimson disc of the sun.

Even seasoned novelists use memetic ticklers to give a good story even more sparkle and more life. Just as an example, renowned science fiction writer Harlan Ellison in *Partners in Wonder* describes how, early in his writing career, he frequently used the device-image of someone shoving his fist in his mouth to demonstrate being overwhelmed by pain or horror. Similarly, S. R. Donaldson, best-selling author of the Thomas Covenant fantasy series, makes frequent use of pet words. For instance, the relatively obscure word "cynosure" (a center of attraction) appears at least once, and usually more often, in most of the books in the series.

Most readers seem to like Donaldson's use of unusual words, though perhaps he teeters close to overdose in such wonderful paragraphs as this one from his book *The One Tree*:

And these were only the nearest entrancements. Other sights abounded: grand statues of water; a pool with its surface woven like an arras; shrubs which flowed through a myriad elegant forms; catenulate sequences of marble, draped from nowhere to nowhere; animals that leaped into the air as birds and drifted down again as snow; swept-wing shapes of malachite flying in gracile curves; sunflowers the size of Giants, with imbricated ophite petals. And everywhere rang the music of bells—cymbals in carillon, chimes wefted into tapestries of tinkling, tones scattered on all sides—the metal-and-crystal language of Elemesnedene.[29]

I just love those "imbricated ophite petals"! Incidentally, H. P. Lovecraft also enjoyed using unfamiliar adjectives like "eldritch," "rugose," "cyclopean," and "squamous."

Antonin Artaud

While still on the topic of words and language, I can't resist mentioning French playwright Antonin Artaud, who once said, "All true language is incomprehensible, like the chatter of a beggar's teeth." I was never 100 percent sure as to what he meant by this, and many people considered him insane. Perhaps Artaud would agree with the observation my friends at work often make: "When people talk, what is actually communicated is often much less than what each party intended to communicate."

Antonin Artaud (1896–1948) was a philosopher and glossolaliac who consumed peyote with the Tarahumara Indians of Mexico. He believed that peyote triggers the brain to remember "supreme truths" difficult to obtain by other means. While using peyote, he saw shapes rise from his belly that looked like the "letters of a very ancient and mysterious alphabet." The letters J and E in particular rose and glowed fiercely.

His final radio play was "To Have Done with the Judgment of God." Artaud wrote this odd piece while in psychiatric institutions, where he was essentially tortured with excessive electroshock and other therapies. He wrote, "I myself spent nine years in an insane asylum, and I never had the obsession of suicide, but I know that each conversation with a psychiatrist, every morning at the time of his visit, made me want to hang myself, realizing that I would not be able to cut his throat."

In 1947, the director of dramatic and literary broadcasts in France commissioned "To Have Done with the Judgment of God," which dealt with consciousness and knowledge, and it was recorded in the final months of 1947. Alas, the broadcast was canceled at the last minute due to its bizarre content, and Artaud dropped dead almost immediately after the cancellation. The "Judgment of God" was not broadcast until the late 1980s.

The radio play ends with a scene in which God lies on an autopsy table. God resembles a body organ removed from the imperfect corpse of humankind. Artaud wanted this closing scene to be superimposed with screams, moans, grunts, snuffling, and glossolalial nonsense words created by Artaud. I do not know if the actual broadcast ended according to Artaud's wishes.

Artaud wrote, "There is in every madman a misunderstood genius whose idea, shining in his head, frightened people, and for whom delirium was the only solution to the strangulation that life had prepared for him."[30]

In the previous chapter, we discussed names of God and the use of Hebrew letters for coding God's name. It is fascinating that people hallucinate letters while using certain drugs, as Artaud did when using peyote. Writer Daniel Pinchbeck also saw letters in one of his LSD-induced psychedelic visions. In particular he saw Hebrew letters "pouring forth from a funnel-shaped geometric mandala."[31]

The Mystery of Italics

Sometimes the Hebrew and other ancient alphabets seen in drug visions are swirly and slanting, like a hyper-italic font. In my books, I try to use italics only sparingly because it is harder to read than nonitalic type. I always thought italics looked rather exotic. Today, italics have invaded every aspect of our lives. However, as I poll passersby on the streets of Shrub Oak, I find that not a single person knows when italics was invented, why it was invented, or by whom it was invented. For virtually everyone alive today, the origin of italics is a mystery.

Let me tear away the veils of this mystery. Italic type, with words slanted *like this*, was invented around 1500 by Italian printer Aldus Manutius. Manutius used italics in a dedication for a book of works by Latin poet Virgil. Manutius's dedication was for his native Italy. The font was based on the cursive handwriting called *Cancelleresca,* used in the government offices of Venice and other Italian city-states. This new slanting style came to be known as Italicus, which means Italian.

The use and meaning of italics has changed through the centuries. For example, words in the King James Bible were often italicized when an editor supplied missing or useful words not existing in the original text to make the text easier to understand. Currently italics are used for emphasizing certain words or phrases, when indicating book or film titles, and to specify words in languages foreign to the language in which a work is written.

Amphigory

My computer poetry, cutup cartoons with incorrect captions, the works of RACTER, and Artaud's writings seem to precariously ride the waves of meaning, close to the edge of insanity and chaos. The term "amphigory" usually refers to a poem or other piece of text that appears to be coherent but, upon closer inspection, contains no meaning! One of the best examples is "Nephelidia," a poem by English Poet Algernon Charles Swinburne (1837–1909). The poem captures the style, alliteration, and rhythms of his other meaningful poems. Here are the first few lines:

"Nephelidia"
From the depth of the dreamy decline of the dawn
through a notable nimbus of nebulous noonshine,
Pallid and pink as the palm of the flag-flower
that flickers with fear of the flies as they float,
Are they looks of our lovers that lustrously lean
from a marvel of mystic miraculous moonshine,
These that we feel in the blood of our blushes
that thicken and threaten with throbs through the throat?
Thicken and thrill as a theatre thronged
at appeal of an actor's appalled agitation,
Fainter with fear of the fires of the future
than pale with the promise of pride in the past;
Flushed with the famishing fullness of fever
that reddens with radiance of rathe recreation,
Gaunt as the ghastliest of glimpses that gleam
through the gloom of the gloaming when ghosts go aghast?

The remainder of "Amphigory" is in Note 32. For some of my own experiments with amphigory, words, books, and the Google Web page, see Note 33.

World's Largest Vocabulary

English has the world's largest vocabulary, with over 800,000 words (including technical words), in part because English borrows words from many other languages. Consider for example the sentence:

> The evil thug loafed beside a crimson table, drinking tea, and eating chocolate, while watching the girl wearing the angora shawl.

Thug is a Hindustani word. *Loafed* is Danish. *Crimson* is from Sanskrit. *Tea* is Chinese. *Chocolate* is Nahuatl. *Angora* is Turkish. *Shawl* is Persian.

Numerous English words were originally French, added to an essentially Germanic language. English also absorbed words from Latin because of the influence of Medieval and Renaissance scholars and the church. Running the world's greatest empire meant that the British were exposed to more other languages than any other imperial power. Harvard Professor Steven Pinker notes that our large vocabulary allows us to be more concise, because a speaker is free to choose a single word with just the right shade of meaning. He suggests that a document in English translated into French is 20 percent longer—"that's the price the French pay for having an academy that keeps out words."[34]

I might add that the number of words needed as we transit between languages also has do with syntax and spelling conventions. In German *Bundestag* is one word, whereas *Federation parliament* in English is two and *chambre de deputés de la fédération* in French is six.

These days, it becomes increasingly difficult to determine the number of words in a language, particularly now that English words are making significant penetration into other languages. English has become the language of international communication and is becoming even more dominant as the Internet unites the planet. Although Chinese is spoken by the most people, English is the most widely spoken second language.

Despite the huge English vocabulary, author Stuart Berg Flexner in *I Hear America Talking* notes that ten simple words account for 25 percent of all English speech, 50 words account for 60 percent, and just 1,500 to 2,000 words account for 99 percent of all that Americans say. Extremely smart and educated people may use as many as 60,000 English words, or about 7.5% of the entire English vocabulary.

The Return of Joumana

I'd like to conclude this smorgasbord of a chapter with a short interview with Joumana Medlej, the Lebanese journalist and designer whom I mentioned in the previous chapter.

CLIFF: Joumana, do you really think that language shapes our perception of reality?

JOUMANA: I believe language creates a structure in our minds through which we perceive reality. Although language was originally shaped to reflect reality, because a single human life is so much shorter than the life of the language, language ensures that the new generations' minds grew within this mold. The process is in truth back-and-forth, humanity shaping language shaping humanity, but on the scale of the individual the only significant influence is that of the language on the person.

CLIFF: Do you think people or cultures with large vocabularies perceive the world differently than those with smaller vocabularies?

JOUMANA: To some extent, yes. It is not that the former see or hear more things than the latter. Rather, having a word for something means there is awareness of its existence as a separate entity. The English language only has the word "love" where the ancient Greeks had "eros," "philos" and "agape." This allowed the Greeks to easily understand three separate emotions, whereas English speakers may sense that it is unnatural to have a single word for a wide panoply of emotions, and compensate with more elaborate descriptions ("I love you like a brother"). In this example, it is not that English speakers can't feel what the Greek did—not at all, it's simply that they don't have the linguistic tools to validate a certain feeling as something experienced by many. That is the primary function of language: connecting a concept in one person's mind to the same concept in another person's, by convention. When a vocabulary is poor, the excluded concepts remain an individual, unshared, undefined experience.

CLIFF: Can you give another example?

JOUMANA: To a significant number of Americans, there exists only one word to designate the populations of the Middle-East—"Arabs."

Consequently, it is very difficult for them to realize that they are not dealing with a single homogeneous bloc where I live, but with a number of very different cultures. Yet today languages are interacting in unique ways and giving people the opportunity to discover and comprehend concepts that don't exist in their own tongue. It's very exciting, even though it means a loss of linguistic innocence.

CLIFF: Do you think Hopi Indians may have a better chance of understanding quantum mechanics (or modern theories of physics that say time is an illusion) than non-Hopis? Why?

JOUMANA: I would rephrase the question as "than certain non-Hopi cultures," because certain other cultures are already equipped to understand the paradox at the heart of quantum physics. Western cultures have issues with it because they leave no room for paradox. Their languages are very keen on fragmenting the world and labeling each part very clearly and unambiguously. The Hopi language, on the other hand, classifies things differently, without necessarily separating them. Their minds are already trained to grasp juxtaposed states that would look irreconcilable to Westerners.

CLIFF: What about the mind of someone who speaks more than one language?

JOUMANA: Being Lebanese, I was raised speaking three very different languages: one Anglo-Saxon, one Latin, and one Semitic. It is very interesting to observe how the early implantation of different patterns in your mind, instead of just one, makes you highly adaptable and resourceful in life in general. I would compare it to having a 64-bit system as opposed to a 4-bit: thinking is much more subtle.

If you listen in on a Lebanese conversation, you'll notice that it is unavoidably conducted in two or three languages at the same time: The speakers unconsciously use the most appropriate word at all times, bringing it in from a foreign language if it doesn't exist in Lebanese so that no gaps are left, no concepts left undefined. This is significant because it means that they are equipped to recognize foreign concepts and when faced with new ones, won't unconsciously try to bend them to a familiar pattern of thinking.[35]

If language and words do shape our thoughts and tickle our neuronal circuits in interesting ways, I sometimes wonder how a child would develop if reared using an "invented" language that was somehow optimized for mind-expansion, emotion, logic, or some other attribute.

Perhaps our current language, which evolved chaotically through the millennia, may not be the most "optimal" language for thinking big thoughts or reasoning beyond the limits of our own intuition. If certain computer languages are more suited for modularity, size, speed, or ease of use, could certain human languages be optimized for human growth potential, creativity, memorability, or for communicating one's thoughts and emotions?

> "No one has ever written, painted, sculpted,
> modeled, built, or invented except
> literally to get out of hell."
> —Antonin Artaud

> "Words fascinate me. They always have. For me, browsing in a dictionary is like being turned loose in a bank."
> —Eddie Cantor, actor (1892–1964)

> "There is a disease which consists in loving words too much. Logophilia first manifests itself in childhood and is, alas, incurable."
> —Peter Ackroyd, "Visions from an addiction to fiction," *The Times* (London), March 20, 2002

> "I have no doubt that our thinking goes on for the most part without the use of signs (words), and, furthermore, largely unconsciously. For how, otherwise, should it happen that sometimes we 'wonder' quite spontaneously about some experience? This 'wondering' appears to occur when an experience comes into conflict with a world of concepts that is already sufficiently fixed within us…The development of the world of thinking is in effect a continual flight from wonder."
> —Albert Einstein, in Paul Schilpp's *Albert Einstein: Philosopher-Scientist*

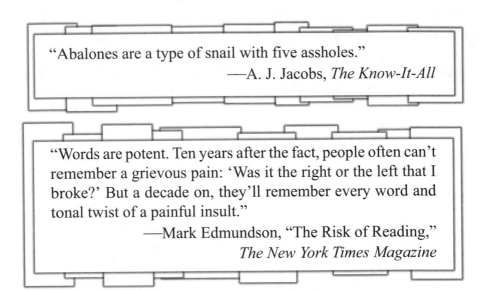

"Abalones are a type of snail with five assholes."

—A. J. Jacobs, *The Know-It-All*

"Words are potent. Ten years after the fact, people often can't remember a grievous pain: 'Was it the right or the left that I broke?' But a decade on, they'll remember every word and tonal twist of a painful insult."

—Mark Edmundson, "The Risk of Reading,"
The New York Times Magazine

DMT, Moses, and the Quest For Transcendence

In which we encounter Humphry Osmond, Aldous and Laura Huxley, DMT machine elves, higher-dimensional dream palaces, the Glass Chrysanthemum, Terence and Dennis McKenna, Rick Strassman, Debra Fadool, reality-enhanced mice, Whitley Strieber, pineal glands, Biblical prophets, ayahuasca, entheogens, Benny Shanon, harmaline alkaloids, Daniel Pinchbeck, robots, larval beings, alien space insects, androids, clowns, dwarves, praying-mantis entities, fairies, cities of Lego with writhing monster squids, Alexander Shulgin, Peter Meyer, Isaac Newton, William James, John Horgan, iboga, Alan Watts, Stanislav Grof, James Kent, near-death experiences, alien senses, the Indian edible-nest swiftlet, "Apparent Communication with Discarnate Entities Induced by Dimethyltryptamine (DMT)," Howard Lotsof, transcendent cable, Ken Wilbur, and temporal lobe epilepsy.

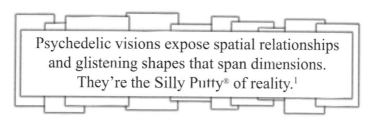

Psychedelic visions expose spatial relationships and glistening shapes that span dimensions. They're the Silly Putty® of reality.[1]

Who Coined the Word "Psychedelic?"

I have no indication that Einstein ever experimented with psychoactive drugs to test their effect on his already hypercreative brain, but, even for Einstein, a feeling of transcendence was important. He said, "I assert that the cosmic religious experience is the strongest and the noblest driving force behind scientific research."[2]

Before we discuss DMT and other psychoactive drugs, I want to remind readers that Marcel Proust also sought "transcendence," a word that Karen Armstrong defines as "the sense of reverence that arises in us when we contemplate the mystery of life," an attitude of awe that springs from that "universal human experience of the numinous....a sense of absorption in a larger, ineffable reality."[3] Proust's "drug" was a madeleine cookie, resembling a little scallop shell. In *In Search of Lost Time*, Proust's mother offers him the madeleine and a cup of lime-blossom tea. He bites into the cookie and a new reality slams into him. He transcends both space and time and is transported as the memories blossom in his mind:

> No sooner had the warm liquid mixed with the crumbs touched my palate than a shudder ran through my whole body. I stopped, intent upon the extraordinary thing that was happening to me. An exquisite pleasure had invaded my senses...with no suggestion of its origin...Suddenly the memory revealed itself. The taste was of a little piece of madeleine which on Sunday mornings my Aunt Leonie used to give me, dipping it first in her own cup of tea....Immediately the old gray house rose up like a stage set—the entire town, with its people and houses, gardens, church, and surroundings sprang into being from my cup of tea.[4]

The new sensations flood Proust as he becomes suddenly indifferent to the usual problems in life and the eventuality of death. He feels an intense sense of love. He said that this mysterious "precious essence" was not in him but rather it *was* him:

> I had ceased now to feel mediocre, contingent, mortal. When could it have come to me, this all-powerful joy? I sensed that it was connected with the taste of the tea and the cake, but that it infinitely transcended those savors...[5]

By some miracle, the cookie that he had not tasted since childhood had the ability to conjure his early days in a town he calls Combray, the old house where his aunt used to live, the water-lilies of the Vivonne, the garden flowers, the country roads, the church, the "good folk of the village and their little dwellings...." Proust wrote that visions took "form and solidity" as they emerged—city and gardens—from his cup of tea.

Food is everywhere in Proust's book: cheese soufflé, a trout with almonds, raspberry mousse....A cookbook exists, entitled *La Cuisine Retrouvee* (by Michéle and Brigitte Gosselin), which contains recipes for every dish mentioned in Proust's work! Other similar books abound and include *Dining with Proust* by Jean-Bernard Naudin, Anne Borrel, and Alain Senderens, and Shirley King's equally captivating *Dining with*

Marcel Proust: A Practical Guide to the Cuisine of the Belle Epoque. Reading groups and even support groups have sprung up to help people get through, appreciate, and share Proust.

The town of Combray, in which Proust's work takes place, is fictional but inspired by the small town of Illiers where Proust actually spent his summers from age six until nine and once again at age 15. Proust seemed to know everyone in the town, and in *In Search of Lost Time*, "Marcel" recognized everyone in Combray, except for the mysterious fisherman whose identity he never discovered. Today, Illiers is called Illiers-Combray to let potential tourists know of its connection to Proust. Pilgrims of Proust flock to Illiers-Combray to visit the local bakeries, eat madeleines, and gaze at the house that Proust called home during the summers.

Today the sky is overcast as I walk down Main Street in Shrub Oak. I can't resist noticing some similarities to Combray. What would Proust have thought of our ancient cemeteries, old trees, and huge mansions next to asphalt streets with modern automobiles? My town was established in 1938—and life ever since then has been nearly idyllic. The schools, the stores…everything's great here. In 1938, a few people who had been renting bungalows and chicken coops in Mohegan Lake got together under the leadership of Irving Breslow to create a "colony" of summer homes, and from this humble start, Shrub Oak grew.

The call of a bird turns my attention to the present. I look up and watch the clouds through the thick, twisted branches of the oaks. I wave to some kids from Lakeland High School and to a few shoppers in the local deli. Today is February 6, 2004—a few hours ago, Humphry Osmond died. Osmond was the British-born psychiatrist who coined the word *psychedelic* to describe the effects of hallucinatory drugs. He also administered LSD and other hallucinogens to healthy volunteers, including doctors, so they could describe their experiences. With that information, Osmond felt that doctors could better understand and care for schizophrenic patients. He also used LSD to treat alcoholics and claimed great success. Unlike Timothy Leary, who sometimes promoted widespread use of psychedelics, Osmond felt that these drugs were "mysterious, dangerous substances that must be treated respectfully."

In the early 1950s, Osmond introduced British writer Aldous Huxley to mescaline, an experience Huxley described in his book *The Doors of Perception.* Osmond first used the word "psychedelic" in a letter to Huxley in 1956. In 1963, when Huxley was dying, he asked his wife Laura Huxley to give him LSD hours before his death.

I first became interested in the mind and psychedelics as portals to new realities after skimming Huxley's *Heaven and Hell*:

> Like the earth of a hundred years ago, our mind still has its darkest Africas, its unmapped Borneos and Amazonian basins. In relation to the fauna of these regions we are not yet zoologists, we are mere naturalists and collectors of specimens. Like the giraffe and the duckbilled platypus, the creatures inhabiting these remoter regions of the mind are exceedingly improbable. Nevertheless they exist, they are facts of observation; and as such, they cannot be ignored by anyone who is honestly trying to understand the world in which he lives.[6]

Huxley also believed that serious individuals might use mescaline or related compounds to grow closer to God:

> To be shaken out of the ruts of ordinary perception, to be shown for a few timeless hours the outer and the inner world, not as they appear to an animal obsessed with survival or to a human being obsessed with words and notions, but as they are apprehended, directly and unconditionally, by Mind at Large—this is an experience of inestimable value to everyone and especially to the intellectual.[7]

This connection to Huxley's Mind at Large motivates primitive cultures to continue to use psychedelics, like peyote and ayahuasca, in their religious ceremonies.

Breaking Through the Glass Chrysanthemum

The possible existence of Huxley's mysterious, timeless regions of reality has led professional and amateur researchers to ponder the extraordinary effects of the DMT molecule (N,N-Dimethyltryptamine). DMT is a psychoactive chemical that causes intense visions and coaxes users to quickly enter a completely different "world" that some have likened to an alien or parallel universe. DMT is a chemical produced in our own bodies and by many plants. Daniel Pinchbeck in *Breaking Open the Head* says, "smoking DMT is like being shot from a cannon into another dimension and returning to this world in less than ten minutes." The transition from our world to the DMT universe ("DMTverse") occurs with no cessation of consciousness or quality of awareness. In DMTverse, beings often emerge and interact with the DMT user. The beings appear to inhabit this parallel realm.

One famous set of DMT studies involves physician Rick Strassman, Clinical Associate Professor of Psychiatry at the University of New Mexico School of Medicine. In 1990, Strassman began the first new research in the United States in more than twenty years on the effects of psychedelic drugs on humans. During the project's five years, he administered approximately four hundred doses of DMT to sixty human volunteers.

We'll discuss the mysterious beings who inhabit the DMTverse in a moment. People using DMT often go through several stages before seeing such creatures. First, these "psychonauts" often see the famous "glass chrysanthemum," circular interlocking patterns in gorgeous colors. Sometimes the chrysanthemum reminds users of a vaulted ceiling or jeweled dome. Users feel as if they are being sucked into a golden funnel filled with exotic characters like runes, hieroglyphics, or Hebrew letters.

For some reason, the concept of language and alphabets are everywhere in the DMTverse. One of Rick Strassman's subjects saw characters resembling a "fantasy alphabet, a cross between runes and Russian or Arabic writing." The subject felt as if information was contained in the writing and that it wasn't random. Another subject saw numbers and alphabetic characters everywhere. Others saw Mayan hieroglyphics.

The glass chrysanthemum comprising overlapping petals of small red, yellow, green and blue rhomboids is a prelude to the DMTverse where the discarnate entities reside. With sufficient doses, the psychonaut is propelled through the chrysanthemum to the DMTverse. The user punches through the membrane of our world and enters a world beyond.

One of my friends who has used DMT says the glass chrysanthemum reminds him of a kaleidoscope. He sees a very well-ordered concentric ring of pulsating, changing colors that completely fills his field of vision. He considered this "the launching pad" to the DMTverse. If the user has a low dose of DMT, that's all he'll ever see. If the user has smoked enough, he will pass through the chrysanthemum phase and "break through" to the land of the elves. My friend says that the post-chrysanthemum realm cannot be objectively explained—it needs to be experienced.

The DMTverse can feel utterly real in terms of detail and potential for exploration. Most subjects have the feeling that it is an independent and constant reality that persists and progresses without their paying attention to it. The reality seems more real than our normal reality. Daniel Pinchbeck, for example, is convinced that the DMTverse is not simply a hallucination. He writes, "I knew it was impossible that my mind, on any level, had created what I was seeing. This was no mental projection. This was not a structure within the brain that the drug had somehow tapped into. It was a nonhuman reality, existing at a deeper level than the physical world."[8] The creatures encountered are often identified as being alienlike or elflike. Some of the creatures appear to be three-dimensional. Others appear to lack depth.

Strassman tried increasing the dose in his subjects. One of his volunteers said "You can still be an atheist until 0.4," referring to a dose of 0.4 mg/kg of body weight. Beyond that, the psychonaut becomes confident that something "real" exists beyond our world.

DMT visions are not just the result of chaotic firing of the eye's rods and cones, in the sense that the psychonaut does not merely experience swatches and swirls of color and disjointed imagery. DMT users often experience closely related and highly detailed themes when in the DMT universe, for example: jeweled cities, jaguars (or other cats), temples, people with gold halos, and objects of gold. These complex objects are frequently seen by users in different settings and cultures. Obviously, these complex visions don't arise from a simple distortion the eye's sensory input from the outside world. Something more complex is going on. But what?

Author and provocateur Terence McKenna has used DMT and feels that "right here and now, one quanta away, there is raging a universe of active intelligence that is transhuman, hyperdimensional, and extremely alien…What is driving religious feeling today is a wish for contact with this other universe." The aliens seen while using DMT present users "with information that is not drawn from the personal history of the individual."[9]

DMT is naturally found in the human body in small quantities. It is also found in many plants, including the Amazonian psychedelic plant brew ayahuasca, also called hoasca (the Portuguese transliteration of ayahuasca) or yagé.

Dr. Rick Strassman, author of *DMT: The Spirit Molecule,* suggests that naturally occurring DMT in the human body may mediate spontaneous psychedelic experiences such as near-death and mystical states. He also believes that DMT is a "reality thermostat, keeping us in a narrow band of awareness so as to ensure our survival." Too little, and our view of the world becomes flat and dim. Too much, and we break through into the machine-elf realm, which I'll discuss in greater detail shortly. We know that DMT is normally found in the base of the spine and in the brain.

Dr. Strassman considers the pineal gland a likely source of this endogenous DMT, in part because many of the necessary ingredients and enzymes for its formation exist in remarkably high concentrations in the pineal gland. He calls the pineal the "spirit gland." Because the pineal gland appears in the developing human fetus around 49 days after conception, perhaps we should not call an embryo truly "human" until DMT production commences. About a century ago, the Catholic church believed that the embryo was not a person until it was 40 days old.

Strassman speculates that DMT, released by the pineal after death, could diffuse to nearby brain areas, allowing the soul to transit. His hypothesis is that the pineal gland produces psychedelic amounts of DMT at extraordinary times in our lives and at death. As we die, our life force leaves the body through the pineal gland, which releases a flood of DMT. Pineal DMT may be produced for a few hours even after we are dead, which could affect any lingering consciousness. The pineal gland is close to the brain's visual, auditory, and emotional centers and thus would be strategically positioned to alter our inner experiences.

Big Wrinkled Simian Brains

When users inject or smoke DMT, or consume ayahuasca, is it possible that the resultant new reality is in some sense a valid reality, on par with our normal reality? Our minds, which, as I've mentioned, evolved to help us run from lions on the African savannas, might not be engineered to see these other realities while in normal biochemical brain states. Certainly, our minds are limited by our brains. Physicists and mathematicians may already be at the edge of possible understanding in such far-flung fields as multidimensional superstrings, parallel universes, loop quantum gravity, motivic cohomology, Langlands' Functoriality Conjecture, large and inaccessible cardinals, and non-abelian reciprocity. Perhaps there is much about the universe we can never see, speak about, or comprehend. Einstein wrote, "I see a pattern, but my imagination cannot picture the maker of that pattern. I see a clock, but I cannot envision the clockmaker. The human mind is unable to conceive of the four dimensions, so how can it conceive of a God, before whom a thousand years and a thousand dimensions are as one?"[10]

Amazingly, the evolution of our big brains may have depended on a single genetic mutation that weakened our jaw muscles about 2.4 million years ago. According to Hansell Stedman of the University of Pennsylvania, the weaker muscles relaxed their hold on the skull, giving the brain room to grow and the skull to expand. Perhaps the weaker jaw place a premium on ingenuity, and natural selection would favor smarter protohumans. Whatever the reason, without that single gene mutation for a particular jaw muscle, humans might not have evolved to be so smart. There would be no Beethoven, Einstein, Saint Peter, or space flight. All of our culture and science is due to a defect in a gene for jaw muscles. Chimps have a genome that is 98% identical to ours, but, unlike us, they still have the gene for producing the special jaw-muscle protein. I imagine them gritting their strong jaws and teeth as they gaze at us with awe and envy.

What is the guarantee that our minds are naturally designed to sense reality as it "really" is? Perhaps there is no guarantee. If this concept seems weird, consider a far-fetched example. Imagine a phenomenon or a creature called a cryptozoid that has been lurking among us since the dawn of evolution. If our ancient ancestors died every time they perceived these phenomena, evolution would favor creatures who did not perceive those creatures or phenomena.

Perhaps DMT is an instrument for seeing the cryptozoic chiaroscuro. As a metaphor, consider infrared goggles. A person leans on a tree. At night, we don't see the person. Put the goggles on, and a new reality results—a truer reality—and we see the man. Similarly, our brain is a filter, and the use of DMT may be like slipping on infrared goggles, allowing us to perceive some aspects of a valid reality that is inches away and all around us. Evolution may act as a valve, preventing us from having too much or too little DMT in our bodies, so that we may survive, dream, and prosper.

Reality-Restricting Genes

As another example of sensory expansion, consider the genes that code the Kv1.3 protein in mice. Researcher Debra Fadool of Florida State University gives mice a super-sense of smell by inactivating the gene for this protein. In other words, the mouse sensorium is held back from its reality perception, until we knock out the reality-restriction gene, and then, suddenly, the mice have a sense of smell 1,000 to 10,000 times as sensitive as that of typical mice. They can now easily discriminate between close odors. A visual analogy is one in which we can suddenly differentiate between close colors of crimson or even see new colors.

So far, Debra Fadool has not found any other physical or behavioral anomalies in mice without this gene. Perhaps in the future, scientists will knock out some of our own reality-limiting genes, take off the brakes, and let us soar. Of course, it may be rather unpleasant if our minds are not able to assimilate the new sensory input, but perhaps we can look for genes that control the quality of information without saturating us with quantity. There is no guarantee that we have evolved to sense reality accurately, or in its many facets. Mental or genetic tampering could make us happier and more creative. Perhaps God applied the brakes by giving us reality-restricting genes until we were sufficiently advanced, like Fadool, to figure out how to remove them.

In my book *The Science of Aliens*, I write about some of Whitley Strieber's experiences with strange realities. Strieber is famous for his book *Communion*, which describes his encounters with aliens. When I had written my book, I felt that Strieber bordered on foolish because some of his descriptions of aliens were zany. Some looked cartoonlike, while others were the more standard versions of aliens that most of us contemplate today. In light of reading about DMT experiences, I can see more of an overlap in Strieber's descriptions and DMT-mediated aliens, elves, and other beings.

We can view alien encounter experiences in a new light, now that we know that DMT can often produce visions of "cartoon-like" aliens and elves. If levels of endogenous DMT in the brain can fluctuate, they could produce such visions. Now, the big question is: is there any actual "validity" to such odd apparitions?

The Glands of Biblical Prophets

DMT in the pineal glands of Biblical prophets may have given God to humanity and let ordinary humans perceive parallel universes. If our human ancestors produced more endogenous DMT than we do today, then certain kinds of visions would have been more likely. Many of the ancient Bible stories describe prophets with DMT-like experiences. Consider as one example (Isaiah 6):

Above him were seraphs, each with six wings: With two wings they covered their faces, with two they covered their feet, and with two they were flying. And they were calling to one another: "Holy, holy, holy is the LORD Almighty; the whole earth is full of his glory." At the sound of their voices the doorposts and thresholds shook and the temple was filled with smoke.

These sorts of ecstatic singing and repetitive exaltations remind me of Rick Strassman's subjects who took DMT and heard: "Now do you see? Now do you see?" along with singing voices. Although speculative, this idea is fodder for a science-fiction novel on which I work. The premise is that the ancients seemed more in touch with God and had more visions than we do today because of endogenous DMT. Maybe Moses, Ezekiel, and Jesus had a greater rate of pineal DMT production than most. Maybe God tampered with Moses's DMT levels to help bridge the gobelin gulf between this world and the world beyond.

When I asked Rick Strassman, M.D., author of *DMT: The Spirit Molecule*, about this theory, he responded:

I'm not familiar with a specific theory that we made more DMT in the old days than now. However, are you familiar with Julian Jaynes' idea of the bicameral mind—that three thousand years ago people routinely hallucinated until the connections between our hemispheres evolved for whatever reason? I don't think hemispheric connections would support his idea as much as higher levels of endogenous DMT might. If indeed we made more DMT in the past, this may have to do with the increase in artificial light that has come upon us in the past 1000 years or so.[11]

I then put the same question to Dennis McKenna, an ethnopharmacologist who has studied plant hallucinogens for more than 20 years. In 1975, he co-authored the book *Invisible Landscape* with his brother Terence McKenna. The book detailed his investigations of Amazonian hallucinogens. He wrote to me:

Cliff, I think it's unlikely that, evolutionarily, we produce less DMT than our ancestors. But I think it's more likely that our ancestors had an intrinsically higher level of pineal activity, because they did not live in an environment with artificial lighting. Light destroys melatonin, and melatonin is the pineal precursor to 5-MeO-DMT and pinoline, a ß-carboline that can shift serotonin metabolism toward the "methylation pathway" when it normally would be degraded by MAO (mono-amine oxidase). One reason modern man is so fucked up, in my opinion, is that we never experience true (absolute) darkness; so the pineal doesn't work properly, we don't sleep properly, and don't have the rich dream life that supports that. So that may be one rationale to support your idea![12]

DMT experiences sometimes include 2-D cartoon-like characters. Often DMT entities lack depth. If DMT levels were higher in ancient humans, would this have affected their artwork? For example, is there anything we can glean from the study of cave paintings and Egyptian art, which tend to lack depth?

Alien Senses

One way to imagine how other realities could exist side by side with our own is to consider the forces that produced the diversity of senses and intelligences right here on Earth. Alien worlds are clearly right here among us. Every Earthly creature perceives the world in an "alien" way. Dogs. Bees. Bats. Cats. They experience the world with different kinds of senses. They can smell what we cannot, they can see what we cannot, they can hear what we cannot. If the organisms of the Earth were somehow able to describe their world, it would not be recognizable to you. It would seem like the wildest world from any science-fiction story. Moreover, if you were able to describe the world to another species, they would "see" no resemblance to their own. We do not have to contemplate science fiction to imagine alienlike senses and bodies. The animal world of Earth is so diverse and full of different senses, that creatures are already walking among us possessing "alien" awarenesses beyond our understanding.

By studying the creatures of the Earth, we get inklings of diverse of realities. Consider the Indian luna moth, which has a wingspread of about 10 cm (about four inches). To our eyes, both the male and female moths are light green and indistinguishable from each other. But the luna moths themselves perceive in the ultraviolet range of light, and to them the female looks quite different from the male. Other creatures have a hard time seeing the moths when they rest on green leaves, but luna moths are not camouflaged to one another because they see each other as brilliantly colored.[13]

The Antipodes of the Mind

Ayahuasca is one entheogen of particular interest to spiritual explorers. The term entheogen comes from *theo*, or god/spirit, and *gen* (create). Thus, entheogens are substances that generate experiences of transcendence and God. Peyote or psilocybin mushrooms are traditional examples of entheogens, and often taken for the purpose of having a mystical experience.

Benny Shanon, Professor of Psychology at the Hebrew University of Jerusalem, has conducted numerous psychological studies of ayahuasca, the plant-based psychotropic brew from the Amazon. The research reported in his book *The Antipodes of the Mind* comes from his own experiences and numerous interviews with people from all walks of life.

The ayahuasca potion is made of two plants, one of which reassembles double-helical vines. Ayahuasca contains the psychoactive chemicals DMT and several harmaline alkaloids. DMT can be extracted and smoked with powerful effect. However, eating DMT does little because MAO (mono-amine oxidase) in our guts deactivates the DMT when DMT is taken orally. On the other hand, the alkaloids in ayahuasca inhibit MAO in the gut, so that the brew is psychoactive. I wonder how the indigenous peoples of the Amazon ever figured out what plants from the thousands of plants in the forest to combine to get both the DMT and the MAO-inhibitor.

Among Shanon's striking findings is that visions different people experience while under the influence of DMT are similar. People with different backgrounds, languages, and cultures often see intricate and majestic landscapes with "lofty towers," factories with precision parts, "steeple-like mitres sewn with precious stones," sparkling palaces with networks of light constructed by "God," and visions that they feel represent the celestial and mathematical fabric on which reality sits. Other authors, like Daniel Pinchbeck, have seen "multidimensional, jewel-faceted places where geometrical and tentacular constructions were being taken apart and reconstructed." One of Rick Strassman's subjects was startled to find herself in a "beautiful domed structure, a virtual Taj Mahal."

According to Kevin Williams, who operates near-death.com, a Web site that is the "Grand Central Station of Near-Death Experiences," people with near-death experiences also often have visions of cities of light—almost identical to those described in the DMTverse. Near-death experiencers describe these cities of light as "golden, beautiful, unearthly, fairytale-like, indescribable, so superior to anything on earth, colorful, brilliant, heavenly, endless, crystalline, grand, paradise, galaxy-like." These sparkling cities "represent an entire world," "radiate with multi-colored lights," are "filled with light beings," "made of glass," "glowing crystals," "castlelike towers," "multidimensional," and "resemble New Jerusalem, the heavenly city in the Book of Revelation." According to Shanon, some of his interviewees said that the ornate DMT palaces were the most beautiful constructs they had ever seen during their lives.

DMT visions closely parallel the jeweled palaces in religious texts describing the afterlife or heaven. For example, the Midrash Konen of the Jews says heaven comprises nested rooms with ceilings of silver, transparent crystal, gold, precious stone, and brilliant lights. Rabbi Joshua ben Levi's midrash describes heaven as having two diamond gates and 600,000 attending angels with shining faces. In the New Testament, the legendary New Jerusalem is constructed from layers of jasper, sapphire, chalcedony, emerald, sardonyx, carnelian, chrysolite, beryl, topaz, chryso-prase, jacinth, and amethyst. The great street of the city resembles gold and sparkling glass.

Many DMT psychonauts experience *presque vu*, or "almost seeing." In other words, psychonauts feel that they are just a curtain away from seeing other figures or divine images. In fact, the entire DMTverse seems to be mere millimeters away from our own. DMT users experience a sense of otherworldliness, of a parallel reality. There is a feeling of enchantment, of sanctity, of beauty—a sense of gaining privileged access to knowledge and intelligence. The world around them appears to be "constructed"—composed with care like a work of art or an intricate hand-spun fabric. Sometimes, people feel they are being guided like actors in a play, and when they look in a mirror, they see their reflections disappear.

Most users feel as if a veil has been lifted, allowing them to view events that have been continuously transpiring in the DMTverse with an existence independent of the psychonaut. The new worlds exhibit internal consistencies and, at times, seems to be even "more real" that our ordinary world. The DMTverse seems to exist independently of our ordinary flow of time.

Many DMT psychonauts return with a *certainty* that consciousness continues after death. In particular, they return to our world believing that their souls exist beyond the body and woven into the fabric of the universe. In John Horgan's *Rational Mysticism*, Stanislav Grof, M.D, Ph.D., explains that his drug use, not confined to DMT, has taken away his fear of death: "Rightly or wrongly, I have the feeling I have been there. I have actually viewed death and thought I died already. I was surprised when I came back. So my present feeling is, it's going to be a fantastic trip. I don't have any sense that it's going to be the end of everything."[14] Grof has observed the effect of LSD on a wide range of people, including the terminally ill whom he has treated with LSD to help them face their impending deaths. He was formerly Chief of Psychiatric Research at the Maryland Psychiatric Research Center and Assistant Professor of Psychiatry at Johns Hopkins University School of Medicine. John Horgan paraphrases Grof: "Consciousness is not merely an epiphenomenon of the brain but can exist as an independent entity."

The U.S. Food and Drug Administration (FDA) recently approved a study to determine if the euphoria and insight of a mild psychedelic experience can ease the physical and emotional pain experienced by terminal cancer patients. Charles Grob, Professor of Psychiatry and Pediatrics at the Harbor-UCLA Medical Center, California, is the lead scientist on a cancer-psilocybin trial. (Psilocybin is a hallucinogenic compound, $C_{12}H_{17}N_2O_4P$, obtained from the mushroom *Psilocybe mexicana.*) During the psilocybin administration, subjects recline wearing eyes mask to screen out distractions while listening to soft music. Perhaps someday the FDA will approve DMT research with similar goals.

The DMTverse is often a curious mix of images and objects present in our world and those native to the DMTverse. For example, DMT psychonauts may see a new world *superimposed* on our "normal world" so that an object from our world is blended with the DMTverse. On the other hand, the psychonaut may see an entirely new world take the place of our world. This new world is navigable as the psychonaut traverses both the new world and our own without bumping into objects such as chairs and walls.

Serpents and cats (like jaguars and pumas) are common in the DMTverse, even if the user is taking a brew in a non-Amazonian setting such as a big city. Black pumas are particularly plentiful. If the user encounters an entity in the DMTverse, sometimes these beings tell psychonauts that they've been "expected." Again, I want to emphasize that people's DMT visions are often very similar, and occur in people from various walks of life, ranging from indigenous Amazonians to New York City dwellers. Unlike dreams, which repeat themes and objects with which the dreamer is typically accustomed, the serpents, temples, gold, and pumas are not common to most psychonauts' experiences in our ordinary world. Shanon states:

> If the commonalties in the ayahuasca experience cannot be accounted for in ordinary psychological terms, then perhaps we have no choice but to shift from the internal domain to the external one and consider the possibility that these commonalties reflect patterns exhibited on another, extra-human realm.[15]

The Inhabitants of the DMTverse Palaces

Daniel Pinchbeck is certain that the DMTverse has a real and separate existence from our own. He writes, "My thesis is that psychedelics actually do reveal other dimensions where there are other forms of beings existing in their own realms. We need to become savvy—eventually, even scientifically precise—about these other realms and develop a kind of protocol for dealing with them."[16] Pinchbeck goes further to say that the DMT beings could be dangerous and uncaring about humanity's interests, making use of humanity in cryptic ways. He worries that our casual use of drugs erodes the boundary between our world and theirs.

I've pored over countless descriptions of the DMT beings. They are frequently described as elflike, alien, or angelic—and of indeterminate gender. Some have multiple faces. Some are made of light. Some seem to materialize from the background and duplicate themselves. A wide range of beings have been described as robots, larval beings, alien space insects, androids, machine-elves, clowns, dwarves, praying-mantis entities, and fairies.

Why so many visions of insects, and particularly of praying-mantis beings? I've come across numerous reports with visions of "lapidary insects," "elflike insects," "cricket creatures," "mechanical insects," "pirate mantids," "mantis beings," "wise mantises," and "mantises in scarlet robes." The naturalist and playwright Maurice Maeterlinck once noted, "Something in the insect seems to be alien to the habits, morals, and psychology of this world, as if it has come from some other planet, more monstrous, more energetic, more insensate, more atrocious, more infernal than our own."[17] Indeed, our insect obsession has penetrated both our dreams and our science-fiction entertainment. We feature insects and bugs in movies like *Five Million Years to Earth* (1948), *Them* (1954), *Tarantula* (1955), *The Deadly Mantis* (1957), *Beginning of the End* (1957), *The Black Scorpion* (1957), *The Naked Jungle* (1954), *Phase IV* (1974),

Empire of the Ants (1977), *Independence Day* (1996), and *Starship Troopers* (1997). Spider beings are prominent in each of Tolkien's three main works, *The Hobbit*, *Lord of the Rings*, and *The Silmarillion*, as he formicates the coruscated depths of our subconscious. In 1897, Fred T. Jane wrote of "big, brainy bugs" on Venus. H. G. Wells populated the moon with insect societies. Giant spiders are drawn in the ancient Nazca lines on the Peruvian desert plateau and on Nazca ceramics. Bushmen worshipped a mantis-god named Cagn or Ikaggen. Scorpion men appear in the *Epic of Gilamesh*. A giant mantis appears in Whitley Strieber's novel *Majestic* and in my novel *The Lobotomy Club*. According to Joe Lewels (*The God Hypothesis: Extraterrestrial Life and Its Implications for Science and Religion*), the God of Moses on Mount Sinai may have been a praying mantis, which is why God did not want to reveal his true face to Moses and humanity. Mantis-alien expert Martin S. Kottmeyer says that alien abductees are now encountering mantis-beings with increasing frequency. These mantis aliens often perform some kind of supervisory role during the abductions.

Everyone seems to have their own perspectives on the DMT beings. Anthropologist Michael Harner met bird-headed people who explained that they were the true gods of our reality. Various Amazonian tribes like the Secoya drink ayahuasca in a deliberate attempt to communicate with heavenly beings. Rick Strassman suggested that the DMT entities actually dwell in the parallel realities posited by modern theories of physics. Terence McKenna called the beings that he frequently encountered "self-transforming machine elves." Like Strassman, McKenna speculated that DMT opens a portal to a parallel reality inhabited by these alien creatures.

Many of the elflike beings remind scholars of entities found in old Celtic tales, which are filled with fairies, elves, and the "wee folk." As recently as a hundred years ago, many people among the Celtic communities of Britain and France believed in the literal existence of these creatures.

Numerous psychonauts have the impression that these entities, sometimes surrounded by complex machinery, are technicians tending and monitoring the very fabric of reality. Daniel Pinchbeck felt as if an "alien intelligence" was coursing through him, examining his nervous system. It was if he were a computer and ayahuasca was "a program performing scans and repairs."

When Pinchbeck studies the effects of ayahuascalike analogs in his home, he listens to music to enhance his experience, particularly: Javanese gamelan (a percussion and gong-chime-dominated musical ensemble), Ravi Shankar, Ornette Coleman, and Bach. While listening to the music during one trip, he suddenly seemed to be on a spaceship. The plantlike creatures piloting the ship "shook their long spilling limbs" at him in greeting.

The palaces and temples in the DMTverse are made out of gold, crystal, or other precious stones. One psychnoaut discussed in Alex Polari de Alverga's *O Livro das Miraçõs* (*The Book of Visions*), describes the palaces: "Thousands of tunnels, galleries, corridors, secret doors, staircases, inclined plane crisscrossed in a composition akin to one of Escher's..." There were "sumptuous palaces, lofty halls, sarcophagi, caverns or temples..."[18] Benny Shanon himself has also frequently seen the palaces and temples of the DMTverse.

O Livro das Miraçõs describes the Santo Daime religion and community, located in the Amazonian rain forest. Members of the religion drink ayahuasca as a shortcut to spiritual transcendence. After years of study, the Brazilian government decided that ayahuasca benefited both society and church members, and declared the drug legal. In fact, the government said that ayahuasca exerted an overwhelmingly positive influence on the lives of users, particularly when taken in a religious setting.

Returning to the DMT palaces, note that entire cities reminiscent of intricate constructions in *Lord of the Rings* are seen in DMTverse, but with a more Chinese or Arabian look. These cities and places are seen by people with no prior expectations about what they would see. Artworks in the cities are made of gold, brass, and crystals.

Vehicles are often seen in the DMTverse, including carriages, boats, and space ships—some ornate and decorated. The vehicles are perhaps metaphors that shout, "Yes! You are on a fantastic journey. Travel with me." Revolving wheels are common as are beings with multiple faces, reminiscent of the multiple-faced beings in Ezekiel 1: 15–22:

Now as I beheld the living creatures, behold one wheel upon the earth by the living creatures, with his four faces. The appearance of the wheels and their work was like unto the color of a beryl: and they four had one likeness: and their appearance and their work was as it were a wheel in the middle of a wheel. When they went, they went upon their four sides: and they turned not when they went. As for their rings, they were so high that they were dreadful; and their rings were full of eyes round about them four. And when the living creatures went, the wheels went by them: and when the living creatures were lifted up from the earth, the wheels were lifted up. Whithersoever the spirit was to go, they went, thither was their spirit to go; and the wheels were lifted up over against them: for the spirit of the living creature was in the wheels. When those went, these went; and when those stood, these stood; and when those were lifted up from the earth, the wheels were lifted up over against them: for the spirit of the living creature was in the wheels. And the likeness of the firmament upon the heads of the living creature was as the color of the terrible crystal, stretched forth over their heads above.

I remind you that Biblical imagery is common in the DMTverse, including Hebrew letters, serpents, and Torahs. Some people see mathematical formulas, which appear to them as conveying basic laws or the mathematical fabric of the universe. Visions of a mathematical fabric are not restricted to ayahuasca users and can take place in a variety of mental states. For example, as I mentioned in this book's Introduction, Indian mathematician Ramanujan dreamed of scrolls containing complicated mathematics unfolding before him. When Ramanujan awakened in the morning, he scribbled only a fraction of what the gods had revealed to him

In some sense, DMT, like Biblical language, primes the mind for transcendent concepts that include angelic beings and temples. But why do so many people see the machine elves, palaces, and cats? Is it possible that the mind has certain worn trenches in which we have a tendency to travel when our blinders are removed, just like the mice with their genes knocked out? Humans seem to be engineered to seek the transcendent and ponder God. Shanon suggests that human beings are also constructed to develop art and worship the holy. If DMT tells us how our minds are architected, perhaps *Homo sapiens* were "engineered" to create palaces and temples, and be delighted by gold and sparkling jewels. Thus, our long history of ornate, inspiring temple building arises because our minds are tuned to create these artifacts. As I expounded in the Preface, just as termites are designed to make intricate mounds, Golden Orb Web spiders to weave tremendous webs, and eagles to create large, intricate nests, so our species is designed to build magnificent jeweled palaces, seek glitter, and compose symphonies.

The pigeon-sized bower bird constructs an ornate nest from twigs, leaves, and moss—and decorates the nest with colorful baubles including feathers, berries, pebbles, and shells. A bird is not taught how to build a nest. Yet each species makes its own style of nest as if it had a built-in blueprint. Bird nests, like human architectures, are quite variable. My favorites are the nests of the "Indian edible-nest swiftlet." The male produces a long, gelatinous strand of congealing saliva, which he winds to create a cuplike nest that gets "glued" a vertical surface. Poachers sometimes steal the glistening, pearly white nests to prepare soups that are said to act as an aphrodisiac.

Cities of Lego with Writhing Monster Squids

DMT is a DEA Schedule 1 substance, which means the government believes it is dangerously open to abuse and has no medical value. Its administration and consumption in the United States are illegal. I am not certain as to the legal implications of drinking ayahuasca, but colleagues tell me that there have been several arrests for possession of ayahuasca or its component plants.

Ayahuasca has now penetrated beyond the Amazon. In fact, it has generated a lucrative "spiritual tourism" industry in Peru. It has been credited with curing thousands of alcoholics and drug addicts. According to most researchers, ayahuasca is not addictive, does not physically harm users, and is unlikely to be widely abused because it often triggers nausea. Journalist Charles Montgomery says, "Church members who drank ayahuasca regularly were more open, optimistic, energetic, without stress, without inhibitions, and had more self-esteem than members of a control group who never drank the tea."[19]

Montgomery himself has tried the ayahuasca brew. He saw "writhing monster squids," "cities of Lego," and "a church and a soaring cathedral, constructed entirely of giant Hallmark greeting cards in pink, blue and pearl." He was never certain if he had made contact with the divine during his ayahuasca journey or was just suffering from chaotic signals in his brain. But he did say that the experience had helped him immensely. "Memories, dreams, anxieties and my own suspicions of the supernatural were somehow transformed into instructive metaphors."

Note that the DMT experience is very different from usual forms of "intoxication." For example, the DMT user often experiences feelings of personal growth and spiritual insight. The drinker has a large measure of control over what is occurring. Users retain their ability to think and reason. Unlike LSD, there are few perceptual distortions in the sense that the images are not melting and twisted. Instead, the user feels as if he or she has contacted a "secret world," something *revealed* that is normally hidden to us. People also hear sounds reminiscent of whining, whirring, chattering, crinkling and crunching—a sound like ripping cellophane. The auditory effects may be dose-dependent. Some users taking small doses hear a distant chanting. With larger doses, people sometimes hear an "elf-language," a rapid chatter that gives users the impression that information is being transmitted at high speed. Terence McKenna in *The Archaic Revival* says, "Nothing prepares one for its crackling, electronic, hyperdimensional, interstellar, extraterrestrial science-fiction quality; it is a complex space filled with highly polished curved surfaces, machines undergoing geometric transformations into beings, and thoughts that condense as visible objects."[20]

Similarly, Shanon in *The Antipodes of the Mind* tells us that the ayahuasca visions are not really like dreams in which the dreamer is the protagonist. Instead, objects—like beings or palaces—seem as if they are revealed to the user to behold, admire, and study. Most dreams involve people the dreamer already knows, and revolve around the dreamer's concerns, present and past—like taking a school test or meeting a friend. DMT visions have nothing to do with the psychonaut's life, and they are not as chaotic or illogical as the plots of our dreams.

DMT also does not simply enhance our capacity for imagination, at least not in the traditional sense. No matter how hard I try, I cannot imagine the vast, intricate, ornate palaces and temples common in the DMTverse, nor can I see them in ordinary dreams. Many psychonauts have the distinct feeling that the visions are not simply products of their own mind. The goal for future researchers is to determine whether the psychonauts can return to our world with new "factual" information. In Chapter 10, I discuss that we should search for these machine elves in ways reminiscent of scientists searching for extraterrestrial intelligence. I realize that any observational programs we design, while solidly based on science, will be on the edge where science meets religion meets science fiction. But whatever scientific searches we develop, they should be based on what we feel might be detectable, regardless of whether we believe the DMT beings to have independent existences from ourselves. We can only hope that the DMT beings can be observed in controlled experiments and that the restrictions of the experiments do not make the beings disappear. If the beings turn out merely to be mental metaphors, let's try to understand why widely diverse groups of people see such similar metaphors. If the palaces and elves are inside us, built into our very nature, we need to know.

Alexander Shulgin once said, "There is a wealth of information built into us...tucked away in the genetic material in every one of our cells. Without some means of access, there is no way even to begin to guess at the extent and quality of what is there. The psychedelic drugs allow exploration of this interior world, and insights into its nature."[21] After receiving his Ph.D. in Biochemistry from the University of California, Berkeley, in 1954, Dr. Shulgin went on to author 200 scientific papers, 20 patents, and 20 book chapters. He has also authored the books *Controlled Substances* (1988), *PIHKAL* (Phenethylamines I Have Known And Loved) (1991), and *TIHKAL* (Tryptamines I Have Known And Loved) (1997), and *Simple Isoquinolines* (2002). Shulgin has dedicated his life to the idea that psychedelics can be used to explore the potential of the human mind. Of the many drugs he has sampled, 2C-T-7 was one of his personal favorites, and you can learn more about 2C-T-7 on the Web. According to people who have experienced this drug, the mental changes it produces are so far beyond the ordinary range of human consciousness that the experiences are as difficult to describe as music would be to a deaf person.

Peter Meyer—mathematician, philosopher, software developer, and DMT researcher—is not sure whether the DMTverse is a different reality, but he is inclined to believe so. He suggests that DMT provides access to an "alternative reality which is full of spirits—the same spirits that shamans encounter." Peter's work is available on the Web and includes a fascinating piece entitled "Apparent Communication with Discarnate Entities Induced by Dimethyltryptamine (DMT)."[22]

Peter's experience with psychedelics such as LSD, psilocybin, ketamine, iboga, harmaline, MDMA, and 2-CB has led him to believe that DMT is the only agent that takes users to the other world "directly, reliably, and quickly." His personal belief is that the DMTverse is an alternative reality that only shamans have explored in a methodical fashion, and he also believes that we will enter DMTverse when we die.

Isaac Newton and DMT

Can we ever understand what the DMTverse is all about? If some aspect of it were "real," could we investigate it and learn from it? When Isaac Newton describes his openness to the paranormal and God, it's almost as if he is speaking of DMT:

> If you ask where the heavenly city is, I answer, I do not know. It becomes not a blind man to talk of colors....But this I say, that as fishes in water ascend and descend, move whether they will and rest where they will, so may Angels and Christ and the Children of the resurrection do in the air and heavens. 'Tis not the place but the state which makes heaven and happiness.[23]

As mentioned in Chapter 1, psychiatrist William James touched on this extra dimension of reality and perception when he wrote, "Our normal waking consciousness, rational consciousness as we call it, is but one special type of consciousness, whilst all about it, parted from it by the filmiest of screens, there lie potential forms of consciousness entirely different....No account of the universe in its totality can be final which leaves these other forms of consciousness quite disregarded."[24]

Despite the remarkable similarities of experiences reported by diverse users of DMT or ayahuasca, my guess is that a user's personality, hopes, and dreams shape the meaning of the experience for the user. The visions of an 18-year-old lifeguard, who has taken ayahuasca at a Haight-Ashbury rave party, will have a different coloration than the DMT visions of a monk studying the Mozarabic Rite in the Capilla Muzárabe Ledo cathedral, or a physicist researching quantum gravity, the physics of time dilation, or Calabi-Yau theory, which suggests we live in a ten-dimensional universe.

Why is DMT a less valid path to God or transcendence than a Gregorian chant, Bahá'í prayer, or Melungeon meditation? How can we say that a mystical experience triggered by prayer is superior in revealing truth and beauty than the experience triggered by drugs or abnormal electrical firing in the brain's temporal lobes? And what about people who both pray and meditate *and* explore with DMT?

John Horgan in *Rational Mysticism* wonders if neurotheologians might one day discover a technology that he calls a "super psychedelic" or "God machine" that can boost the brain's production of endogenous DMT. If technology existed that could safely induce blissful mystical experiences, what effect would this have on the future of civilization? Rick Strassman envisions a genetically engineered virus that turns on some of our methylating enzymes and boosts our internal DMT production. If this virus spread through society, the Horgan "God machine" becomes a possibility within this century.

Iboga and Beyond

DMT is not the only entheogen that promotes encounters with discarnate entities. Iboga, made from African rootbark, can also give users visions of entities. Daniel Pinchbeck, for example, tried iboga and saw a "dark and faceless golemlike figure formed out of rough logs." Crossing its legs, it leaned forward as if curious, studying Pinchbeck. To Pinchbeck, the being seemed "utterly real." Shamans encouraged him to communicate with beings he encountered because they had messages for him. Pinchbeck writes in *Breaking Open the Head* that he had the impression of contact "with some other intelligence or entity existing in a realm outside our own." In contrast to DMT, the ibogaine experience seems to deal more with the personal concerns of the user.

Howard Lotsof, famous for his serious studies of ibogaine, suggests it provides a cure for heroin and cocaine addiction. A former heroin addict himself, Lotsof first obtained ibogaine from a drug researcher cleaning out his refrigerator. Back in the 1960s, he took the drug with friends for recreational reasons. While high and crossing a street, he saw "seven copies of himself, freeze-framed, crossing the street behind him." When Lotsof and his friends' trips ended, virtually all of them were shocked to find they had no desire to resume using heroin. The ibogaine removed his dependency on heroin and cocaine without the pain of withdrawal—a claim endorsed by other addicts who have tried it.

Two decades later, Lotsof patented the iboga molecule under the name "ibogaine" for purposes of addiction treatment, but the FDA refused to approve it. Ibogaine was subsequently declared, along with LSD, an illegal Schedule 1 substance, with potential for abuse and no medical value. The iboga bark's psychedelic power is produced by several alkaloids that appear to affect many of the brain's known neurotransmitters, including serotonin and dopamine. In fact, ibogaine, like LSD, mescaline (the essential alkaloid in the peyote cactus), and psilocybin, has a molecular structure quite similar to the tryptamine neurotransmitter serotonin. The DMT molecule, incidentally, only differs from serotonin by having two methyl groups. Some users say that iboga crams a decade of psychoanalysis into a single day.

Transcendent Cable

I've come to think of the brain as a cable TV box. When an individual has normal levels of neurotransmitters like serotonin, the person has "Basic Cable" akin to the inexpensive service that provides a limited set of TV channels, without HBO and Showtime. By taking DMT, which may affect serotonin levels or substitute for serotonin, the individual has just bought the Premium package with all sorts of new channels. The TV hasn't changed, but it is now able to access new programs. DMT users say that they are not intoxicated, but are simply now able to perceive new broadcasts. As with cable TV, some of the new content can be intense, disturbing, or even dangerous to an inexperienced or young individual.

Both ayahuasca and ibogaine contain natural MAO inhibitors; as a result, the mind-altering compounds are not inactivated so quickly. No one is sure *why* the plants create the psychoactive compounds. Perhaps the chemicals keep away insects or plant-eating animals. Perhaps they attract plant eaters who seek mind-alteration and thus eat the plant and spread the seeds in their feces. People who have just one strong dose of ibogaine can go for many weeks with very little sleep, and scientists wonder how a drug that exits the system rather quickly can exert such a long-term effect.

Personal Mail from DMT Users

I have received numerous letters and e-mails from people who have described their personal experiences of the DMTverse. For example, Roger E. said he saw an entity that had brilliantly colored moving segments all over its skin. The vision seemed entirely real as the entity grinned and exhibited swirling appendages. Roger also sensed the presence of smaller invisible beings probing his body. Roger considers this DMT encounter the most profoundly moving experience of his life.

Olly H. wrote to me that he has experimented with DPT, an analogue of DMT. He encountered elves, which though "good-natured, seemed to have a decidedly mischievous sense of humor." When he tried to focus on the elves, they would fade away, but if he relaxed and didn't make a strenuous effort to see them, they would rematerialize. He writes, "These

DPT elves seemed to find my futile attempts to get a good focus on them rather amusing; they seemed to be saying, 'Catch us if you can!'"

One friend said that he participated in a DMT group excursion in which several people linked hands and had a guide verbally take them on a trip. The entire dream team went back to the beginning of the universe and together experienced the formation of planets and other celestial bodies, and experienced the birth of humanity. He also had a *dream* of a DMT experience that seemed to access the same "doors" as the original DMT experience. Perhaps once the brain is primed, subsequent trips to the DMTverse can be made simply through dreams. My friend says,

> I experimented with DMT on several occasions, and found that it provided access to a truly extraordinary mindspace, characterized by the preposterous but nevertheless wholly convincing sensation that the DMTverse was far closer to "reality as it really is" than ordinary, everyday experience. I was left with the curious impression that everyday experience might more aptly be characterized as the hallucination here, and that the "warped" impression of reality I experienced under the influence of DMT was in fact more "genuine."[25]

One of my colleagues, whom I'll call "Trypt," tells me that the beings he encountered in his living room were "beings of light." He met the creatures in 1996 while using DMT, but remembers the encounter with extreme clarity. Four entities materialized as outlines of standard human forms. Inside each outline was light. The beings seemed quite real to him, more real than any dream he had ever had. The beings communicated through telepathy and told him not to be scared. They understood that this was his initiation, and they were celebrating and rejoicing with him. They encouraged him to continue "seeking."

Trypt felt as if he were a radio signal being tuned in on an AM radio station, and the experience was accompanied by sounds reminiscent of tuning a radio station dial. As soon as he was "tuned in," he was in another place. What happened next was reminiscent of a typical alien abduction story. He was sprawled out on a table, in some alien world, and looking up at two beings. One being said that Trypt was not ready yet, and the other agreed. He was then thrust back into his living room. The experience has had a profound personal impact on the way he views the universe and his life.

Another colleague, Christopher B., experimented with DMT and regularly interacted with the DMT beings. He tells me that they appeared in several forms, were friendly and loving, and answered his questions telepathically. He believes that "these beings are a reflection of reality or how reality would appear if we lost all of our filters." During one trip to the DMTverse, Chris saw the patterns on the walls of his home become coated with "visible language describing everything." This language looked like Sanskrit to him.

> The beings were everywhere and clearly were the underlying construct of everything. They looked like imps, fairies, gnomes, dwarves, extraterrestrials. I call them intra-terrestrials as they were clearly "in" everything. They certainly were the "wiring under the floor boards" as the late Terence McKenna described in his experiences of the same space that is intimate in the instruction of this reality.[26]

What are we to make of my colleagues' experiences? Do they tell us about the workings of reality, the mind, or both? Alan Watts (1915–1973), who held both a master's degree in theology and a doctorate of divinity, once wrote:

> There is no difference in principle between sharpening perception with an external instrument, such as a microscope, and sharpening it with an internal instrument, such as one of these…drugs. If they are an affront to the dignity of the mind, the microscope is an affront to the dignity of the eye and the telephone to the dignity of the ear. Strictly speaking, these drugs do not impart wisdom at all, any more than the microscope alone gives knowledge. They provide the raw materials of wisdom, and are useful to the extent that the individual can integrate what they reveal into the whole pattern of his behavior and the whole system of his knowledge.[27]

A Word from "Skeptic" James Kent

James Kent wrote to me about his own feelings regarding DMT. Kent is the former editor of *Psychedelic Illumination* magazine, former publisher of *Trip Magazine*, and is now curator of tripzine.com. He told me that he has frequently experimented with DMT in the past, has interviewed and spent time with the late Terence McKenna, is friends with Rick Strassman, and has studied DMT very closely for the past fifteen years. He personally does not believe DMT is a gateway to an alternate dimension, nor that it provides literal contact with elves and alien entities. He concurs that DMT can produce a vivid, otherworldly landscape, often including elves, aliens, insects, snakes, and jaguars. However, he believes the DMTverse to be no more than an aberration of the brain's perceptual mechanics. According to Kent, when DMT enters the brain, perceptual distortion occurs in the same way that a line of code inserted into a computer program can drastically alter the way information is presented on a computer screen. He also points out that the sensation of seeing aliens, elves, or being in the presence of God(s) is not unique to DMT users. Otherwise sane people who have never tried DMT report these sensations all the time, and it is generally treated as a sign of psychosis. Recent research has shown that, by stimulating parts of the temporal lobe, one can reliably reproduce the feeling of being in the presence of God or angels, or of being watched.[29]

Kent has seen the elves in his DMT experiences (and on Psilocybe mushrooms as well), and did indeed perceive them as externalized disincarnate beings, even managing to carry on rudimentary conversations. However, the more he experimented with DMT, the more he felt that the elves were mere machinations of his mind. While under the influence, he found he could think them into existence, and then think them right out of existence at will.

Why is the alien/elf archetype so common to the DMT experience? Kent believes that we have innate evolutionary wetworking that forces us to latch onto any piece of anthropomorphic data that pops up in otherwise random sensory data, such as spotting a face peering out from behind the bushes, or spotting another human form hiding in the tall grass. The evolutionary advantage of such a trait is obvious. In standard Rorschach tests, even the most ambiguous blobs are found to look like faces or people no matter what culture the observer is from. Given the amazing swirling kaleidoscopic imagery produced in the typical DMT trip, it is inevitable that anthropomorphic shapes will emerge. For Kent, the appearance of elves is less interesting than the messages they sometimes give, which Kent says revolve around the environment or botanical ecosystems. Even the skeptic Kent wonders if DMT provides some way for "plant consciousness to commune with us."

DMT, Mystical Truth, and Temporal Lobe Epilepsy

Can drugs like DMT lead to quick philosophical or mystical truths? On one hand, Ken Wilbur, one of the world's most widely published philosophers, says that spiritual truths require hard work:

> If you want to see a cell nucleus, look down this microscope. On the spiritual side, if you want to see your Buddha nature, if you want to see Christ consciousness, if you want to see the religious side of the equation, fold your legs, sit down each day for two hours, count your breath from one to ten. Do that for five years and get back to me.[28]

On the other hand, this difficulty of achieving mystical states has led people like Terence McKenna to conclude that DMT provides a faster and more reliable route. He says, "Psychedelics are democratic. They work for Joe Ordinary. I can't go and sweep the ashram for eighteen years or some rigmarole like that."[29]

Researchers Andrew Newberg and Eugene D'Aquili discovered that mystical experiences, in which one loses all sense of self, corresponded to suppressed activity in a particular part of the brain. In particular, for most Buddhists practicing Tibetan meditation and nuns in contemplative prayer, activity decreases in the posterior superior parietal lobe, which helps us orient our bodies as we traverse the world. In John Horgan's *Rational Mysticism*, Newberg says that mystical experiences should be taken seriously because they often "feel truer, more *real*, than…ordinary perception; they seem to represent a more fundamental reality than the baseline reality."[30] Newberg suggests that our seeking mystical experiences may have evolved because they often produce better interpersonal relationships and a positive outlook on life.

Various religious prophets such as Saint Paul, Ezekiel, and Mohammed may have suffered from epileptic seizures. Recently, several nuns with temporal lobe epilepsy (TLE) have provided evidence that TLE is the root of many mystical religious experiences. For example, one former nun "apprehended" God in TLE seizures and described the experience:

> Suddenly everything comes together in a moment—everything adds up, and you're flooded with a sense of joy, and you're just about to grasp it, and then you lose it and you crawl in to an attack. It's easy to

see how, in a prescientific age, an epileptic or any temporal lobe fringe experience like that could be thought to be God Himself.[31]

Even the Old Testament prophet Ezekiel had a TLE-like vision reminiscent of modern UFO reports—the famous, fearsome Ma'aseh Merkabah, the Vision of the Chariot:

> And I looked, and behold, a whirlwind came out of the north, a great cloud, and a fire enfolding itself, and a brightness was about it, and out of the midst thereof as the color of amber, out of the midst of the fire....Also out of the midst thereof, came the likeness of four living creatures. And this was their appearance, they had the likeness of a man. And every one had four faces, and every one had four wings. And their feet were straight feet; and the sole of their feet was like the sole of a calf's foot; and they sparkled like the color of burnished brass.[32]

Eve LaPlante, author of *Seized*, is just one of a growing number of writers and researchers delving into TLE-induced religious experiences. For example, Professor Michael Persinger from Laurentian University in Ontario, Canada, researches the neurophysiology of religious feelings, and believes that spiritual experiences come from altered electrical activity in the brain. David Bear from Harvard Medical School believes that "a temporal lobe focus in superior individuals (like van Gogh, Dostoevsky, Mohammad, Saint Paul and Moses) may spark an extraordinary search for the entity we alternatively call truth or beauty." Religion, then, is sometimes our interpretation of altered temporolimbic electrical activity. This is not to demean the mystical experience, because TLE personalities have obviously accomplished great things, whose depth and meaning have radiated far beyond the electric storms of a single cranium.

LaPlante, in her book *Seized*, aptly sums up the growing evidence linking TLE and creativity:

> Hidden or diagnosed, admitted or unknown, the mental states that occur in TLE seizures are more than simply neurological symptoms. In people like Tennyson, Saint Paul, and van Gogh, these states may have provided material for religion and art. People with TLE, whether or not they know the physiological cause of their seizures, often

incorporate their symptoms into poems, stories and myths. And the disorder does more than provide the stuff of religious experience and creative work. TLE is associated with personality change even when seizures are not occurring; it amplifies the very traits that draw people to religion and art.[33]

As for me, the bulk of my mind stimulation and mystic transport comes from the *Encyclopaedia Britannica* and its wonderful panoply of topics that induce neural nirvana. Writer A. J. Jacobs expresses the pleasure and shocking mind-expansion one receives from the *Britannica*, as he completes his quest to read all entries in alphabetical order:

> The changes are so abrupt and relentless, you can't help but get mental whiplash. You go from depressing to uplifting, from tiny to cosmic, from ancient to modern. There's no segue…Just a little white space, and boom, you've switched from theology to worm behavior. But I don't mind. Bring on the whiplash—the odder the juxtapositions, the better. That's the way reality is—a bizarre, jumbled-up Cobb salad.[34]

"That's the thing I like best about this ayahuasca tea. For hundreds of years, mainstream Christians have been expected to communicate with God through intermediaries of our churches. Our God, it seemed, stopped addressing us personally 2,000 years ago. Well, the ayahuasca-based churches cut out the middlemen."
— Charles Montgomery,
"Think LSD and Ecstasy Have Devoted Followings?"[35]

"If God takes DMT, does He see people?"

"I had the impression of bursting into a space inhabited by merry elfin, self-transforming, machine creatures. Dozens of these friendly fractal entities, looking like self-dribbling Fabergé eggs on the rebound, had surrounded me…babbling in a visible and five-dimensional form of Ecstatic Nostratic."
— Terence McKenna, *True Hallucinations*

"The ephemeral nature of peak experiences sparked by psychedelics makes them no different from any other sort of mystic encounter with the *mysterium tremendum*. Such soul-rocking events are indelible in spite of their transient nature, whether you're a born-again Christian or an acid mystic turned Buddhist monk. But the degree to which they will affect you over time, and the tenacity of your newfound conviction, depend on how well you integrate the often alien or otherly vision into your daily life."

—Charles Hayes,
"Is Taking a Psychedelic an Act of Sedition?" *Tikkun*

"Scientist Stephen J. Gould researched and wrote his critically acclaimed 1,400-page *The Structure of Evolutionary Theory* while frequently using marijuana to maintain his health. He writes: "The most important effect upon my eventual cure [relief from nausea due to cancer chemotherapy] was the illegal drug, marijuana...Marijuana worked like a charm. It is beyond my comprehension that any humane person would withhold such a beneficial substance from people in such great need simply because others use it for different purposes."

—Stephen J. Gould[36]

"Was it translinguistic matter, the living opalescent excrescence of the alchemical abyss of hyperspace, something generated by the sex act performed under such crazy conditions?"

—Terence McKenna, *True Hallucinations*

Brain Syndromes Open Portals to Parallel Universes

In which we encounter the illusion of intermetamorphosis, cryonics, delusional parasitosis, perception puzzles, doppelgängers, asparagus, bipolar disorder, Kay Jamison, creativity, Special K, angels, Irish Fairy and Folktales, Invasion of the Body Snatchers, and psychiatric disorders such as Charles Bonnet syndrome, Capgras syndrome, Fregoli syndrome, asomatognosia, prosopagnosia, reduplicative paramnesia, the syndrome of subjective doubles, Cotard's syndrome, and Ekbom's syndrome.

A mind is an endless train weaving its way
through the landscape of reality.
But who made the train tracks,
and where is the conductor?[1]

Proust and Asparagus

A big rubber brain sits atop my computer monitor at work. I squeeze it from time to time to reduce stress. Sometimes I imagine myself as a brain-snatcher—the Chief Curator of a futuristic collection of famous brains. The brains are stored on shelves in a cavern, beneath the basement of my home. I walk down lamp-lit corridors filled with gray, wrinkled things stored in formalin-filled jars to prevent decay. Stalagmites support some of the shelves. The coolness of the cavern contributes to the brains' remarkable state of preservation.

Did you know that a significantly large number of established writers and artists have had bipolar disorder, a disease in which patients oscillate between depression and hyperactive euphoria? In my imaginary collection, there are plenty of artists' brains. Follow me. Look toward your left. Here are the brains of the brilliant writers, artists, and composers who had bipolar disorder: Sylvia Plath, Walt Whitman, Cole Porter, Anne

Sexton, Vincent van Gogh, Gustav Mahler, John Berryman, Edgar Allan Poe, Virginia Woolf, Herman Hesse, Mark Rothko, Mark Twain, Charles Mingus, Tennessee Williams, Georgia O'Keefe, and Ezra Pound. In one smaller bottle are some fragments of Ernest Hemingway's manic-depressive brain—all that is left after he blew his brains out. (Bipolar disorder used to be called manic depression.)

I give a little tap on the jar marked "Poe." I lift the jar and tap again. His cerebrum jiggles like a nervous prune. *Nevermore, nevermore.*

As I point out in my book, *Strange Brains and Genius: The Secret Lives of Eccentric Geniuses and Madmen*, genius and insanity are often entwined.

I put Poe back in his place. In the distance I hear a low moan, but it is probably just the movement of wind within some of the narrower corridors.

Professor Kay Redfield Jamison, in her *Scientific American* article "Manic-Depressive Illness and Creativity," clearly demonstrates that established artists have a remarkably high incidence of bipolar disorder or major depression.[2] Established artists and writers experience up to 18 times the rate of suicide seen in the general population, 10 times the rate of depression, and 10 to 20 times the rate of bipolar disorder. During periods of mania, people with bipolar disorder have sharpened and unusually creative thinking, increased productivity, expansive thoughts, and grandiose moods. They can overcome writing blocks, generate new ideas, and have better performances. People with bipolar disorder rhyme more often and use alliteration more often than unaffected individuals. They also use idiosyncratic words three times as often as control subjects and can list synonyms more rapidly than normal. As with temporal lobe epilepsy, some patients with bipolar disorder stop taking their medications because the drugs can dampen their emotional and perceptual range as well as their general intellect.

I take a few steps northward in my cavern of brains. The cave walls are reddish, glittering with garnet and quartz crystals. The air smells clean and wet. Huddled in concentric rings are outcroppings of aragonite, clustered near a clear pool. The larger crystals remind me of fingers on a huge hand. Above the crystals hangs a large photo of Kay Jamison.

Jamison—who is also a psychiatrist at Johns Hopkins University School of Medicine in Baltimore and author of *Touched with Fire: Manic-Depressive Illness and the Artistic Temperament*—suggests that mania can be conducive to creativity because it allows afflicted individuals to work long hours without sleep, to focus intensely, to experience a variety of emotions, and to make bold assertions without fear of the consequences. Bipolars take risks and are open to contradictions. Jamison's descriptions about the effects of bipolar disease are sometimes poetic and mystical. In 1995, Jasmison revealed in her memoir *An Unquiet Mind* that she, herself, has bipolar disorder. She describes a preternatural awareness of surroundings, a sensitivity to the environment that is more animal-like than human. She says, "You can't predict what you'll be like tomorrow. I think the moods of bipolarity mirror the natural world, which is so seasonal and fluid. It's a dangerous amphibious sort of existence."[3]

I walk a little farther, wrinkling my nose at the strange chemical odors that suddenly flood the tunnel. Perhaps one of the jars is open.

In the future, no one will have to worry about the negative side of bipolar disorder because we will have found the genes responsible for the debilitating aspect of the afflictions, which should allow us to dampen the problematic symptoms while allowing individuals to retain their creative spark.

On my right are a few clear pickle jars. I reach for the one marked "Marcel Proust," open it, and let my fingers linger over his gray-white frontal lobes. Might there be remnants of his genius preserved in his neuronal networks? Proust was tormented by depression as well as a host of other ailments such as asthma, dizziness, and fainting spells. Given this, he nevertheless persevered to write *In Search of Lost Time* before he died at age 51.

The brain: three pounds of soft matter that can take a split second of experience and freeze it forever in its cellular connections. A hundred billion nerve cells are the architecture of our experience. Is Proust still here in the wet organ balanced in my palm? Could we reconstruct his memories? Would Proust approve such a breach of privacy?

My cryonicist friends refuse to give up hope that memories still reside in the brain cell interconnections and chemistry, much of which can be preserved. Maybe they are right. After all, far back in the 1950s, hamster brains were partially frozen and revived by British researcher Audrey Smith.[4] If hamster brains can function after being frozen, why can't ours? In the 1960s, Japanese researcher Isamu Suda froze cat brains for a month and then thawed them. Some brain activity persisted.[5]

But what if there is an afterlife? I bang on the bottle containing Proust's brain, causing the brain to make a splashing sound like a drunken fish. If there is an afterlife, he must have already experienced it by now. What would happen if his brain were revived?

I suspect that Marcel Proust's brain was different from the average brain. Proust's memory modules and sensory apparatuses were somehow more acute than the rest of ours. His realities, like the neorealities I describe in Chapter 8, were alive with detail, existing outside of space and time. As just one example of this hyperreality, look at Proust's magnificent description and memory of his boyhood fascination with asparagus:

> My greatest pleasure was the asparagus, bathed in ultramarine and pink and whose spears, delicately brushed in mauve and azure, fade imperceptibly to the base of the stalk—still soiled with the earth of their bed—through iridescences that are not of this world. It seemed to me that these celestial nuances betrayed the delicious creatures that had amused themselves by becoming vegetables and which, through the disguise of their firm, edible flesh, gave a glimpse in these dawn-born colors, these rainbow sketches, this extinction of blue evenings, of the precious essence that I would still recognize when, all night following a dinner where I had eaten them, they played in their crude, poetic farces, like one of Shakespeare's fairies, at changing my chamberpot into a bottle of perfume.[6]

To Proust, every moment, every object, and every emotion became an opportunity for exuberant introspection. Proust's town of Combray, like my own Shrub Oak, is the kind of small town where each new day is like the previous, and the smallest departure from routine is cause for notice. Socrates once said, "The unexamined life is not worth living." Proust took this philosophy to the limit, becoming hyperSocratic.

Even the trees spoke to Proust in *Within a Budding Grove*: "I saw the trees drawing away, waving their desperate arms, seeming to say to me: 'What you don't learn from us today, you'll never know. If you let us sink into the depths of this road from which we'd sought to raise ourselves to you, a whole part of yourself that we were bringing to you will forever fall into nothingness.'"[7]

Perception Puzzle Interlude

Before launching into neurological syndromes that alter human realities in the most remarkable ways, I would first like to introduce you to a puzzle I recently invented. I am forever obsessed with how we perceive reality, so I ask you this: How many squares can you count in this window below? I designed and created this stained-glass window in gorgeous color, and my brainy colleagues always give different answers as to the number of squares! Show this to friends. No two friends will give you the same answer as this puzzle seems to play with our perception of consensus reality.

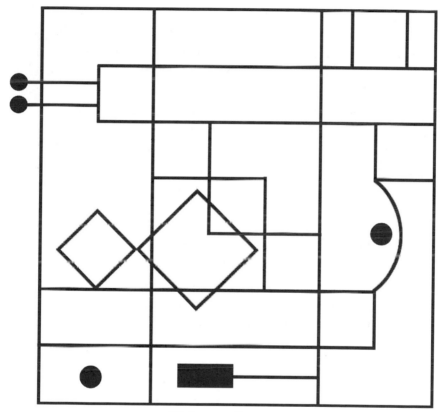

Pickover Square
How Many Squares Do You Count?

How many squares did you find? If you are a teacher, how many squares can your students find? How can our perceptual systems be so different? How does the number of perceived squares depend on a person's age, gender, occupation, emotional state, IQ, caffeine level, hypnotic state, and culture?

Bonnet People: Welcome to a New Reality

For a long time, I have been interested in brain syndromes that appear to open portals to new realities. For example, in the previous chapter, we discussed temporal lobe epilepsy and its effect on religiosity and transcendent experience.

Here I want to focus on people afflicted with certain eye diseases who report encounters with elflike beings from parallel universes. In particular, people with Charles Bonnet syndrome see beings from another world. Many scientists would call these beings hallucinations. Others call this syndrome a portal to a parallel reality. These individuals with Charles Bonnet syndrome (or "Bonnet people") are otherwise mentally sound. The beings appear when Bonnet people's visions deteriorate as a result of eye diseases such as age-related macular degeneration—or when patients have had both eyes removed. Charles Bonnet syndrome is more common in older people with a high level of education.

Bonnet people report that they see apparitions resembling distorted faces, costumed figures, ghosts, gouls, fairies, witches, little people, and beings wearing large hats, looking like they just walked out of Tolkien's *Lord of the Rings*.

Dr. Stephen Doyle, an ophthalmologist at Manchester Royal Eye Hospital, and Robert Teunisse, a psychiatrist at the University Hospital in Nijmegen, Netherlands, describe a very sane woman who was sitting quietly at home when she suddenly saw several "two-inch-high, stovepipe-hat-wearing chimney sweeps parading in front of her."[8] She tried to catch one but could not. Her only medical problem was that she had poor sight due to macular degeneration.

Dominic Ffytche, of the Institute of Psychiatry in London, has studied many Bonnet people. One of his patients described how a friend working in front of a hedge suddenly disappeared, as if behind an imperceptible veil. "There was an orange peaked cap bobbing around in front of the hedge and floating in space by its own devices."[9]

Interestingly, Daniel Pinchbeck—while experimenting with mushrooms and moclobemide (an antidepressant containing a mono-amine oxidase inhibitor to enhance the tryptamines in the mushroom)—saw a linear configuration of laughing green elves wearing peaked caps. Pinchbeck says that these elves were totally realistic, as clear to his inner vision as film images. Why do drugs and brain syndromes continually bring us back to elves that also appear to permeate the folklore of many cultures?

According to Ffytche, 50 percent of Bonnet-people see a disembodied or distorted face of a stranger with staring eyes and prominent teeth. Sometimes the strangers are seen only in an outline or cartoon-type form, which reminds me of some images seen by people taking DMT. Occasionally, the faces are grotesque and resemble the faces of gargoyles. Some people see beings with multiple faces, such as those seen by people using DMT.

Beings in the Bonnet realm sometimes have blank eye sockets. This image is also occasionally reported by people who have used the hallucinogen Special K (ketamine hydrochloride). One person e-mailed me to tell me that, while under the influence of Special K, everything was normal except that people in the room suddenly had no eye sockets, just black voids, and he saw light being sucked into the void from around the periphery of the eyeballs. Ketamine has also been associated with feelings of ego loss, out-of-body experiences, and near-death experiences.

Bonnet people also sometimes see serene landscapes or vortices. Many see entire new worlds, replete with landscapes or groups of people, which are either life size or tiny. Perhaps when vision deteriorates, the brain's visual cortex is starved for information, and the brain is free to access parallel realities or create new ones.

Researchers at the Low Vision Unit of the Department of Ophthalmology, University Hospital, Nijmegen studied hundreds of Bonnet people and conclude that the imagery in their visions can be complex, almost comical—for example, "two miniature policemen guiding a midget villain to a tiny prison van."[10] Others perceived beings that included ghostly translucent figures floating in hallways, a person wearing a large flower on his head, and shining angels. As with people taking DMT, the Bonnet universe is often seen with the subject's eyes open. Like the DMT experience, the objects superimpose well with the rest of the real world—for example, an elf sitting on a real couch or tree stump. As with DMT, the "Bonnetverse" is not personally relevant to the subject. For example, the subject does not see friends and family materialize in the Bonnetverse. Most Bonnet people do not tell physicians about their visions, fearing that they will be considered mentally ill.

Bonnet Syndrome is named after the Swiss philosopher Charles Bonnet, who first described this condition in the 1760s. Bonnet's grandfather, who was blinded by cataracts, suddenly saw birds and buildings that Charles Bonnet could not see. I believe that Charles Bonnet may have later experienced the hallucinations himself as his eyesight declined.

We already discussed in Chapter 4 the possibility that the brain was tuned or primed for certain kinds of visions, which DMT helped reveal. In our daily lives, the images we "see" are determined by external sensory inputs and internal inputs, such as when we visualize an apple or an elf. Normally, our external sensory input is sufficient to crowd out the internal visions. However, Bonnet people may have so little visual input that the internal images begin to emerge. It's as if a whole cabal of witches, dwarfs, ghosts, and elves are swimming around in our minds all the time, and it takes Bonnet syndrome to let them punch through into our consciousness. Dreams have a similar genesis in that they are triggered internally, often from memories.

My colleague James Kent would like to see how people with Charles Bonnet syndrome react to Pindolol, which is a blocker for 5HT2A, a molecular brain receptor. Pindolol administration can terminate the DMTverse immediately. What effect would this have on the Bonnetverse?

Capgras Syndrome

Imagine that you are sleeping. It is seven AM. With unnerving suddenness, the alarm from your electronic clock wakes you from the depths of a pleasant dream. "Damn," you whisper. A romantic image fades like an angel drifting off into distant clouds. Why does the alarm have to go off now?

Even though you'd had the clock since high school, you've never become accustomed to its shrill beeping. Within seconds you silence it by reaching for a button on its surface. Your finger encounters the old familiar scratch from the previous time you threw the clock across your bedroom. Now the scratch allows you to guide your finger to the button bringing blessed silence.

You slowly sit up in bed, push aside your blankets, and turn on *The Today Show* with your TV's remote control. You listen with half an ear as Katie Couric interviews Madonna. Outside, a few birds chirp in the cloudy morning air.

There's just one problem. A big problem. You look around. Your bedroom is a cheap facade. The bird outside is a mechanical contrivance that flies away screeching. Someone has taken away all your family members and replaced them with duplicates. The new family members look and act the same as your original family. They can share your memories and laugh with you and reminisce. Yet, in spite of the close similarity, you "know" your "new" spouse is not your original spouse. Even your pet has been replaced with a nearly identical replica. Welcome to the world of Capgras syndrome.

People with Capgras syndrome ("Capgras people") act as if they are in a parallel universe in which family members are "doubles" or "impostors." When Capgras people see a friend, spouse, or themselves in a mirror, they believe they are seeing an impostor who looks just like the original person. Sometimes Capgras people even believe that inanimate objects—like a chair, watch, book, or lamp—have been replaced by exact replicas. If Capgras people own a pet, the pet may be seen as an impostor, a strange animal roaming through their lives and homes. In one case, a man with a large pituitary tumor believed that his wife had replaced hundreds of his possessions with similar, but inferior replicas.

Capgras patients are often so disturbed when they see their *doppelgängers* in the mirror that they remove all mirrors from the home. (*Doppelgänger* is the German word for a ghostly double of a living person.) Capgras syndrome, named for French psychiatrist Jean Marie Joseph Capgras, afflicts thousands of people in the United States. Some people with Capgras syndrome have epilepsy or strange-looking temporal lobes in the brain. In fact, about 30% of Capgras cases are associated with obvious brain pathologies like those produced by head traumas or epilepsy. However, these conspicuous physical factors do not seem necessary or sufficient to explain the peculiar manifestation of Capgras' syndrome. Some Capgras people exhibit an imbalance in dopamine or serotonin levels in the brain.

What generalizations can we make? An analysis of dozens of Capgras people suggests that, in most married patients, the spouse is the primary double. In widows, other relatives are the doubles. In single persons, usually the parent or sibling becomes the double. The Capgras patient identifies his or her spouse as being a fake—identical in every possible way, but still a replica. The patient will accept living with these impostors but will secretly "know" that they are not the people they claim to be. This reminds me of the movie *Invasion of the Body Snatchers*.

Fregoli Syndrome, Asomatognosia, and Beyond

Many syndromes related to Capgras come to mind. Consider Fregoli syndrome, in which the patient insists that someone who is actually unfamiliar to her is someone whom the patient really knows—sort of a reverse Capgras. Sometimes strangers' bodies are thought to hold the minds of loved ones or enemies. A Fregoliac might also believe that one or more familiar persons are persecutors following the patient and repeatedly changing their appearances. The condition is named after the Italian actor Leopoldo Fregoli (1867–1936) who was famous for his ability to make quick changes of appearance during his stage act.

I've also had friends with relatives who were asomatognosiacs. People with *asomatognosia* do not recognize a portion of their bodies as their own. You might point to an asomatognosiac's hand, ask them what it is, and the asomatognosiac might say "my brother's hand." People afflicted with *prosopagnosia* are unable to recognize faces, even if they can recognize almost everything else. Prosopagnosiacs can be quite intelligent and live fairly successful lives, but they can walk past family members at a party and not recognize them. Often a prosopagnosiac tries to identify and distinguish family members by studying their clothing. Anecdotes exist of a prosopagnosiac at a party who offends a woman by smiling at her and saying, "Come on, dear, let's go home." The woman was *not* his wife.

M. David Enoch and Hadrian Ball list several other related syndromes in their blockbuster book *Uncommon Psychiatric Syndromes*:[11]

- *Reduplicative paramnesia*, in which patients believe that a physical location has been duplicated (described by Picko in 1903).
- *The illusion of intermetamorphosis*, in which the patient believes that people he knows change with one another. For example, his friend Joe becomes Bill. Bill becomes Sam. Sam becomes Joe (described by Courbon and Tusques in 1932).
- *Syndrome of Subjective Doubles*, in which the patient believes that other people are becoming him. In some cases, the patient sees doubles of himself morphing into other people, or the patient believes himself to be an impostor or in the process of being replaced. This syndrome was described by Siomopoulos and Goldsmith in 1975.

Cotard's and Ekbom's Syndromes

People with Cotard's syndrome mistakenly believe that they have lost organs or that they have died and are walking corpses. Jules Cotard first described the syndrome in 1880 and noted that these people sometimes believe that their brains, stomachs, hearts, or spirits are missing. The person feels damned and that his body is reduced to a machine. At times, Cotard people feel that their bodies are infinite in extent. Some believe they have turned to stone or that their bodies are just artificial shells.

Cotard's first case of this condition involved a 43-year-old woman who believed that she had "no brain, nerves, chest or entrails." She also believed, like many Cotard people, that she was "eternal" and would "live forever." Charles Bonnet, the French physician whom we already discussed, described a woman who, in 1788, insisted on being placed in a coffin, in which she stayed for weeks. She believed herself to be already dead.

Many other fascinating cases are discussed in the previously mentioned classic clinical guide *Uncommon Psychiatric Syndromes*. For example, one woman said that the Earth, Sun, and other stars did not exist, and that she alone survived the Big Bang. She believed that she wandered an empty world in a form of a "carbonized star." She also said that time ceased to exist and that she was condemned to wander for eternity.

A man with Cotard's syndrome once asked to be buried because he already considered himself to be a corpse, and had no flesh, legs, or torso. Another patient said he was a "living corpse with no stomach and bowels." Yet another said he had "no blood and no veins." One woman continued to believe that something offensive was emanating from her rectum, a symptom that reminds me of Ekbom's syndrome, also known as "delusional parasitosis."

In Ekbom's syndrome, people are certain that they are infested with insects or worms. Swedish neurologist Karl Axel Ekbom (1907–1977) described the syndrome in 1937, and today we know that the typical patient is a woman over 40 years of age. One 68-year-old woman in the literature continually complained about "something crawling up my private." She insisted that worms had invaded her rectum and spine. She searched for these for weeks and said she saw them in her feces. This behavior is typical in that most women imagine that the parasites are infesting their outer rectal areas. They frequently feel that bugs, worms, or mites are burrowing into, under, or out of the skin.

Another woman, for a decade, believed that insects crawled through her hair. She felt them and said they made "cracking noises." Insects are sometimes "seen" emerging from cosmetics or toothpaste. Some women forge physician signatures and spend their entire savings on prescription drugs that can be used to treat insect or worm infestation. They chemically burn themselves with various agents in an attempt to kill the parasites. The patient is also often compelled to dig the parasites out using knives or tweezers, leaving behind skin lesions.

People with Ekbom's syndrome often bring samples of the supposed infestation to physicians, such as an envelope filled with particles, dust, skin flakes and other material. Microscopic examination has always confirmed that of the contents of the envelopes consist only of lint, scabs, or other household dust.

DMT, Bonnet Syndrome, and Beyond

Let's wrap up Chapters 4 and 5, both of which touched on mental anomalies. As we have discussed, people using DMT or with Charles Bonnet syndrome see elves, aliens, and other beings. The idea that these kinds of beings are behind our obvious reality—mending it, tending it, constructing it, and sculpting it—has probably been with humankind since emergence from our apelike ancestors. In 1888, poet William Butler Yeats (1865–1939) wrote in *Irish Fairy and Folktales*:

> Behind the visible are chains on chains of conscious beings, who have no inherent form, but change according to their whim, or the mind that sees them. You cannot lift your hand without influencing and being influenced by hordes. The visible world is merely their skin. In dreams we go amongst them. They are, perhaps, human souls in the crucible—these creatures of whim.

And yet I continue to wonder how a few molecules of DMT or even macular degeneration can cause us to imagine such complex and recurrent themes, like elves with hats or vast landscapes. With DMT, there are the space beings, crystal palaces, pumas, and metal machines. Although Chapter 4 focused on DMT, other drugs, such as DPT (N,N-dipropyltryptamine and longer acting than DMT), also appear to reveal intelligent entities. It's as if the elves are all around us. Many cultures have legends of demonic beings who torment us while we lie helplessly in bed at night—such as succubi and incubi in medieval Europe. Centuries before, in Greece, people contemplated nature spirits such as the Alseids, Auloniads, Dryads, Hamadryads, and Meliae. The Poles and Lithuanians had their Wila, shape-shifting souls of the dead. The Slavs had their Vila, nymphlike creatures who controlled storms, and the female ghost Rusalka. The Norse had their Askefruer, who dwelt in the trees, and the British had Jenny Greenteeth, the scary water hag, living in the treacherous river trees England. Given the common reports of aliens, UFOs, ghosts, spirits, angels, Ieimakids, Oreads, and Eleionomae in the last few decades, perhaps the same regions of the brain that are activated during the DMT or Charles Bonnet experiences can account for an even wider phenomenon, a trans-Potameides consciousness, that pervades all cultures for all time.

"And what is the dimension beyond life as illuminated by DMT? It seems to be a nearby realm inhabited by eternal elfin entelochies made entirely of information and joyous self-expression. The afterlife is more Celtic fairyland than existential nonentity..."
—Terence McKenna, quoted in *Breaking Open the Head*[12]

"...The tool of psychedelic consciousness is certainly not an imperative, and not for everyone; it must be utilized, managed, and regulated skillfully. In order to fill the sensorium with as much preternatural light as can be metabolized, and liberate the psychedelic experience from the underworld darkness of proscription, the practice should be sacramentalized and institutionalized under the administration of the scientists, doctors, psychologists, and spiritual leaders most knowledgeable about its propensities and potentials."
—Charles Hayes,
"Is Taking a Psychedelic an Act of Sedition?" *Tikkun*

"I think, on an unconscious level, we're more in tune with the realities of existence than we are on a conscious level...Our culture narrows our vision. It allows us to see things in a certain way, but prevents us from seeing in other ways. We may have the capability to see three hundred sixty degrees, but we basically see life in a narrow wedge. If only we could look at the world with no prejudices, with our senses wide open to reception, we might see everything in a totally different way..."

—Dean Koontz in *Dean Koontz: A Writer's Biography*[13]

"Patients [with Ekbom's syndrome] most commonly present to dermatologists (if skin changes are present) or pest control officers (if they are not)....The patient is compelled to dig the parasites out, especially before going to bed, and often resorts to the use of a knife, tweezers or other sharp implement, leaving skin lesions consistent therewith....The symptoms may be carried over, or passed on by suggestion, to other members of the family."

—Dr. Phillip Weinstein, "Insects in Psychiatry,"
Digest of Cultural Entomology

"The [visions] were of the same character as the images of the kaleidoscope, symmetrical groupings of spiked objects. Then, in the course of the evening, they became...a vast field of golden jewels, studded with red and green stones....Then they would spring up into flower-like shapes beneath my gaze, and then seem to turn into gorgeous butterfly forms or endless folds of glistening, iridescent, fibrous wings of wonderful insects."

—Henry Havelock Ellis,
"Mescal: A New Artificial Paradise"[14]

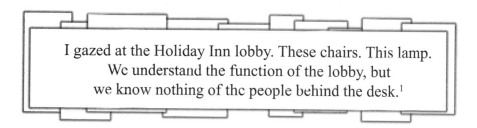

From Holiday Inn to the Head of Christ

In which we discuss quantum immortality, psychedelic Shakespeare, Marcel Proust, the Meseglise and Guermantes ways, Dr. Brown's Cel-Ray soda, Holiday Inn founder Kemmons Wilson, TV laugh tracks, Thomas Jefferson's decimated Bible, Warner Sallman's Head of Christ, Mel Gibson, Lucia Joyce, Carl Jung, Agent 488, General Dwight Eisenhower, Walter Benjamin and the Arcades Project, Tolkien's Lothlorien, the chanting of Sausalito mystery fish, the tenacity of the Acoemeti, recipes for attracting women, word salad and computer spam, dropping pennies from the Empire State Building, movie closing credits, near-death experiences, living in the π matrix, parallel universes, the many-worlds interpretation of quantum mechanics, Robert Heinlein, The Wishing Project, John Goddard, and the mathematical abilities of Jesus.

I gazed at the Holiday Inn lobby. These chairs. This lamp.
We understand the function of the lobby, but
we know nothing of the people behind the desk.[1]

According to Roger Shattuck, author of *Proust's Way*, reading Proust is like visiting a zoo: "The specimens Proust collected from the remotest corners of society amaze and amuse us in their variety....[and] roam free in a vast park." I hope the specimen topics in the current chapter challenge you with their color and variety. Consider the following pages as a stream of consciousness, an impressionistic collage of disjoint subjects that personally excite me. Relax. Breathe deep. Welcome to the Reality Carnival.

The Meseglise Way and Guermantes Way

In *In Search of Lost Time*, the character "Marcel" reminisces about his early years spent with relatives in the town of Combray. At one end of his aunt's house is a door that leads to a walking path called Meseglise Way, also called Swann's Way. The other leads to Guermantes Way. On one level, they are just paths that traverse the village and on which Marcel's family takes daily walks. One path goes to the estate of the wealthy Guermantes family, the other to Swann's middle-class estate. However, they represented much more to Proust—different directions in life and the choices we make. At the end of his masterpiece, the narrator, who has grown old, revisits Combray and discovers a shortcut that unites the two paths. He realizes now that the two "ways" are connected after all:

> Thus for me, do the Meseglise Way and the Guermantes Way remain linked to so many small events of that one life of all the diverse lives that we lead on parallel lines, the one which is the fullest of events, the most rich in episodes, the life of the mind.[2]

Although the Guermantes Way leads to the elegant chateau of the aristocratic Guermantes family, Proust never actually seems to reach the chateau because the walking distance is too great. Thus, one path represents a path to the ordinary, and the other represents a path to the furthest reaches of space, time, and mind.

When I walk through my own town, I try to discover paths that link apparently different "ways." In my writing, I always endeavor to link seemingly disparate branches of human activity, and this is also why I write on so many different subjects, from black holes to medical mysteries to God.

I'm thirsty now. Let's begin our stroll in Shrub Oak and find something to drink.

Dr. Brown's Cel-Ray Soda

Over the past twenty years, I've seen Shrub Oak in Yorktown change. Its northern edge, near Route 6, was once quite rural but is now jammed with traffic. Many of the old wooded lots have been cleared to make way for new stores. Most of the progress has been good. I love the Jefferson Valley Mall, which is the setting for one of my science-fiction novels, and I enjoy the new Barnes & Noble along Route 6.

Modern technology is now everywhere in Yorktown. Some of the hip new bar owners contemplate selling drinks that glow under ultraviolet light. This idea is not far fetched. Don Keiller, a chemist at the Anglia Polytechnic University in Cambridge, U.K., has recently developed fluorescent proteins, harvested from algae. One of the chemicals, a phycoerythrin, glows a lurid luminescent pink in an alcoholic drink exposed to ultraviolet (UV) light. Alas, people can't drink the stuff until scientists prove it's safe to drink.

Personally, I rarely drink alcoholic beverages. The most exotic beverage I routinely imbibe is a Doctor Brown's Cel-Ray Soda, along with a corned beef sandwich at Bloom's Kosher Deli in downtown Yorktown. The deli is next to a pet store, where I buy food for my tropical fish, and Hanada Sushi, where I eat spicy tekka maki. I'm sipping on the Cel-Ray soda right now.

Does celery-flavored soda seem odd to you? Have you ever tried it? It used to be quite popular, with many dozens of brands made since the 1880s. There was celery tonic, celery cream, celery cola, and celery beer. However, I believe that the only celery-flavored soda sold today is Dr. Brown's Cel-Ray. The label on the bottle says, "Since 1868."

According to Cel-Ray soda historian Dennis Smith, Dr. Brown was either a beverage chemist, a doctor treating immigrant children on Manhattan's Lower East Side, or a work of fiction.[3] First called "Dr. Brown's Celery Tonic," the name was changed to Cel-Ray around 1930. In those days, Dr. Brown's Cel-Ray Tonic was sold in every Jewish deli in New York, and the drink contained celery seeds and sugar. In the early thirties, before Coca-Cola became kosher, many Jews started drinking Cel-Ray soda as well as Dr. Brown's cream and cherry sodas.

Kemmons Wilson, Holiday Inn, Nirvana

All this talk of Cel-Ray soda brings a flood of memories from my childhood, when I ate at all the local Jewish delis near Asbury Park, New Jersey, with my parents. Kosher pickles. Matzoh ball soup. Corned beef. Hot pastrami. One of my favorite delis was opposite a small Holiday Inn with its 43-foot green-and-yellow sign surrounded by glowing neon marquee lights. I always wanted to venture inside that Holiday Inn with its blinking lights, but my parents never seemed interested.

Kemmons Wilson (1913–2003) founded Holiday Inn, the wholesome chain of roadside motels that took its name from a Bing Crosby movie. When they first opened, these hermitic bubbles percolated in a sea of dangerous uncertainty and allowed humankind (well, at least people with cars) to venture beyond the confines of their homes and pierce their own little portions of the cosmic egg.

In the 1930s and 1940s, the American highways were lined with dirty diners and fleabag lodges populated by truckers and whores. Thanks to Wilson's vision, the highway gloom began to sparkle as familiar antiseptic chrome and clean beds triumphed over dirt and danger, and the darkness gave way to a clean cultural phosphorescence.

Wilson was already a millionaire by 1951 from his chain of popcorn machines, real estate deals, and a jukebox franchise. At the time, he became unhappy with the state of motels in America. He wanted a chain of motels that was safe and clean, and where children could stay without charge.

Wilson's first Holiday Inn opened in 1952. Each room had a private bathroom, air conditioning, and a telephone. The hotel had a swimming pool, free ice cubes, free parking, and dog kennels—all pretty revolutionary for the early 1950s. Ten years later, the number of Inns had multiplied to a walloping 400. At the peak of its growth, a new Holiday Inn was opening somewhere in the world every two and a half days. Holiday Inn signs lit up skylines from Texarkana to Tahiti, from Hong Kong to Casablanca.

The Holiday Inn was an experiment in cultural technology, making it easy for most people to take long-distance vacations. It helped to change behavior on a large scale, which *New York Times* writer Walter Kirn says became a ritual "whose manifold consequences are unknowable."[4]

To revive my childhood feelings, I recently drove to the nearby Holiday Inn in Mount Kisco, New York, sat in the lobby, and smiled at the people behind the desk. I sighed, staring at the Inn's finely detailed pine furnishings, huge stone fireplace, and the Hudson River mural…This seemed to be fancier than the Inns I recalled from childhood. I imagined Kemmons Wilson smiling back at me.

Wilson's "be here now" philosophy is exemplified by one of his guiding principles for success, espoused in his autobiography *Half Luck and Half Brains*: "You can't change the past, but you sure can ruin the present by worrying over the future. Remember that half the things we worry about never happen, and the other half are going to happen anyway. So, why worry?" Because the exuberant Kemmons Wilson propelled suburban pioneers into the wilderness of the great Interstate unknown, he can rightly be called the Timothy Leary of highway travel.

Holiday Inn let people experience novelty in their lives, safely, just as Rick Strassman did in his controlled, hospital experiments with DMT, which we discussed in Chapter 4. Holiday Inn introduced "free will" to the masses. The Interstate highways beckoned. Everyone could become an explorer.

Laughter: Mind Bend or Mind Control?

One of the ultimate "experimental drugs" in the later part of the twentieth century was the TV laugh track. Charles Douglas (1910–2003) invented the laugh track for television comedies. His 1950s invention, called the "Laff Box," had a mechanism that he jealously guarded, never revealing just how it worked. Researchers have since confirmed that laugh tracks increase the likelihood that a viewer will consider a situation funny. We laugh in order to fit into and conform with our social surroundings. Professor Robert Provine, a psychology professor at the University of Maryland, speculates that we have a laugh-detection circuit in our brains, which is triggered when we hear others laugh. Douglas has shown us that our emotions and predilections are easily sculpted. We are not in control. We are sheep.[5]

The first television show featuring a laugh track was NBC's *The Hank McCune Show* (1950). *The Abbott and Costello Show* used a laugh track that ran continuously, regardless of the action on screen. According to laugh-track expert Ben Glenn II, Glen Glenn Sound perfected the laugh track process in the 1960s, and their tracks and engineers dominated the industry. Thus, the same laugh tracks can be heard on nearly every sitcom of the era, independent of the TV studio producing the show. In the 1960s, sound technicians used a two-foot-tall apparatus with ten horizontal and four vertical keys and a foot pedal. The engineer "orchestrates" the laugh track by using the keyboard to select attributes such as gender and type, while using the foot pedal to determine laughter duration.

The concept of canned laughter goes back to the 4th century BC, when Greek playwrights hired audience members to laugh at their comedies. In the 19th century, most theaters in France hired sets of people to laugh and clap. Each person had a specific function. For example, the *bisseurs* shouted for encores; the *rieurs* laughed at comedies, and the *pleureuses* wept at tragedies.

These days, when I study laugh track sounds, I sometimes hear an "anticipatory" chuckle, perhaps meant to imply that some bright person has seen the joke coming before everyone else, and then there is a ripple of laughter from others, like waves produced by a pebble thrown into a pond.

Several years ago, writer Paul Krassner (Chapter 3) studied humor in situation comedies. He suggests that continued exposure to "canned laughter" reprograms the nervous system. "Now, when something on TV struck me as funny," he writes, "I noticed myself hesitating for just a split second—waiting for permission from the laugh track. I had lost my instinctive sense of critical judgment."[6]

I've not come across "antilaugh" tracks that insert gasps in order to instill fear in scary movies. It seems that music usually performs this function. In *Jaws*, the music "shrieks" just as a head floats into view, presumably prompting the audience to do likewise. In Alfred Hitchcock's *Pyscho*, we hear "shrieking" violins. In pornographic movies, we have the various moans, with voices that are sometimes not the actor's voices, or at least not synchronized with the on-screen action, but added to excite the viewer. One of my colleagues compares laugh tracks and other canned sounds to sympathetic vibrations and the resonances of bells. People ring like a bell when they hear laughter, even without the presence of something funny. Perhaps the laughter response also suppresses critical filters, which is why salesmen and politicians tell jokes.

Thomas Jefferson Uses Razor on Bible

The laugh track machine provided a way to insert extra content in broadcasts in order to shape the viewer's reaction to the ersatz reality portrayed on TV or radio. More than a century earlier, Thomas Jefferson was doing just the opposite—removing content from the most famous presentation of all time in order to understand its essence when stripped of its emotion.

In 1804, Thomas Jefferson used a razor to remove all passages of the King James Version of the New Testament that had supernatural content such as the virgin birth, resurrection, or turning water into wine. About one-tenth of the bible remained, which he pasted together and published as *The Philosophy of Jesus of Nazareth*. Apparently, Jefferson admired Jesus as a teacher and prophet but was not always interested in the cloak of divinity. Thomas Jefferson wrote, "Of all the systems of morality, ancient or modern, which have come under my observation, none appear to me so pure as that of Jesus. [*The Jefferson Bible*] is a paradigma of the doctrines of Jesus, made by cutting the texts out of the book and arranging them on the pages of a blank book, in a certain order of time or subject.

A more beautiful or precious morsel of ethics I have never seen."[7] He told John Adams that he excised "pure principles which Jesus taught," from the "artificial vestments in which they have been muffled by priests, who have travestied them into various forms as instruments of riches and power for themselves."

Jefferson was disappointed by the seeming injustices in the traditional Bible. He wrote to clergyman Ezra Stiles Ely (1786–1819):

> I am of a sect by myself, as far as I know. I am not a Jew, and therefore do not adopt their theology, which supposes the god of infinite justice to punish the sins of the fathers upon their children, unto the 3rd and 4th generation.[8]

Jefferson's skepticism is reinforced by his letter to John Adams:

> The whole history of these books [the Gospels] is so defective and doubtful that it seems vain to attempt minute inquiry into it: and such tricks have been played with their text...that we have a right, from that cause, to entertain much doubt what parts of them are genuine. In the New Testament there is internal evidence that parts of it have proceeded from an extraordinary man; and that other parts are of the fabric of very inferior minds. It is as easy to separate those parts, as to pick out diamonds from dunghills.[9]

Finally, Jefferson reflects on the mystery of Jesus in a later letter to John Adams:

> And the day will come when the mystical generation of Jesus, by the supreme being as his father in the womb of a virgin will be classed with the fable of the generation of Minerva in the brain of Jupiter. But may we hope that the dawn of reason and freedom of thought in these United States will do away with this artificial scaffolding, and restore to us the primitive and genuine doctrines of this most venerated reformer of human errors.[10]

Jefferson's Bible is an example of many expurgated versions published through the centuries. In 1796, the Bishop of London created a Bible in which chapters were rated to indicate violent and pornographic content. In the 1800s, the Quakers released a Bible with the indecent material toned down and printed in italics so that could be avoided. Various "family Bible" publishers today have taken a razor to the "offensive" parts.

Albert Einstein might have agreed with many of Jefferson's ideas and the fractured-and-sliced Bible. In 1933, Einstein wrote, "If one purges all subsequent additions from the original teachings of the Prophets and Christianity, especially those of the priests, one is left with a pedagogy that is capable of curing all the social ills of humankind."[11] In 1943, Einstein reaffirmed, "It is quite possible that we can do greater things than Jesus, for what is written in the Bible about him is poetically embellished."[12]

Fond Memories of Warner Sallman's Head of Christ

When I was a child growing up in Middletown, New Jersey, I was the only Jewish kid in the neighborhood and entire school. After second grade, however, we moved from Middletown to Ocean Township, and suddenly Jews were everywhere. Before our move, I recall that portraits of Jesus were ubiquitous in neighbors' homes. I asked one little friend what Jesus was all about, and he replied, "He watches to make sure we brush our teeth."

I also remember that no other image of Jesus was etched so deeply into my mind as the ubiquitous "Head of Christ," painted in 1940 by Warner Sallman (1892–1968). This bearded Jesus has hung in suburban kitchens, bedrooms, dens, and hallways across America ever since. If you don't know to what image I refer, go to a computer and Google "Sallman." The portrait is everywhere. The image has been reproduced in countless media, including special Bibles, church literature, calendars, posters, buttons, and bumper stickers. The Sallman Christ has been reproduced over a billion times. More than any other religious icon, this single image propelled Jesus from an abstract man to living flesh. It is the *de facto* and authentic portrait of Jesus for generations of many Christians.

According to some friends, the painting was so moving that their religiosity increased as a result of viewing it, and it became the centerpiece of their faith. The remarkable influence of the Sallman Christ prompted the Lilly Endowment to fund a major study of the impact of the painting on the American psyche. Many women surveyed thought that the image was an accurate rendition of the historical Jesus. In the painting, Jesus wears a robe and has long hair and a beard. Nothing in the painting's background suggests where Jesus is standing. The lighting is soft, almost as if Jesus posed in a photography studio for the photographer. Some people pray to the photo. Others claim that it weeps, bleeds, or oozes oil.

Warner Sallman showed a talent for painting at an early age and was encouraged by his artistic parents. He was mesmerized by religious art, such as stained-glass windows and paintings of Biblical scenes. Legend has it that Sallman's Head of Christ came to Sallman in a dream. Shrewd marketing by the Gospel Trumpet Company and the start of World War II encouraged many American soldiers to carry card-sized prints of the Head. Sales skyrocketed.

These days, art critics often see the painting as overly sentimental and corny. Jesus is too white, too effeminate, too clean, too pure. According to James Elkin, author of *Why Are Our Pictures Puzzles?*, viewers saw all kinds of "hidden figures" in the painting, even those Sallman never consciously intended. Sallman nevertheless incorporated them in his lectures, highlighting what his fans found with chalk: a wafer on Jesus's forehead, a chalice on his temple, a cross and dove under his right eye, and a nun, monk, and Jewish prophet on the right shoulder. In subsequent paintings, Sallman intentionally inserted such hidden images.

In the 21st century, what new image might overtake the image of the Sallman head? Many people see an accurate rendition of Christ in actor James Caviezel's portrayal of Jesus in Mel Gibson's movie *The Passion of the Christ*. A few fans of the movie bow down before Caviezel who said that he was the same age as Jesus when Jesus was crucified (33 years old) and that the actor's initials, J. C., are suggestive of Jesus Christ.

During the first several hundred years of church history, people were not driven to draw and sculpt Jesus, as far as I can tell. Later, the question of Jesus's face inspired artists from Michelangelo and Rembrandt to Dalí. Humanity will not rest until it has a clearer picture of the son of God.

The December 2002 *Popular Mechanics* featured a painting of Jesus's face that many scholars think is more accurate than Sallman's. This Jesus has a broad face, dark olive skin, short curly hair, and a prominent nose. To create this face, forensic anthropologists and computer programmers started with an Israeli skull dating back to the 1st century. They then used computer programs, clay, computer-generated skin, and an array of physical-anthropological knowledge about ancient Jewish people to determine the shape and color of the face. The researchers said it was unlikely that Jesus had long hair, given Paul's statement in First Corinthians, "If a man has long hair, it is a disgrace to him." However, this statement has many interpretations; for example, Paul may have said this to assure the gentile Christians that they did not have to act like the typical Jewish Christians of the day.

Note that the researchers did not put clay directly on the ancient skull because rabbinical laws prohibit tampering with any part of the skeleton. Instead, researchers performed a CT scan of the skull and then constructed a replica for their use.

Lucia Joyce, Mad Daughter

For a long time I've been haunted by the dizzying image of Lucia Joyce (1907–1982), the beautiful and mad daughter of James Joyce and Nora Barnacle. In the early 1930s, Lucia began to act erratically—setting fires, singing all night, sending telegrams to dead people, and going into trances. When friends called James to congratulate him on winning his obscenity trial in the United States, thereby enabling the publication of *Ulysses*, Lucia cut the phone wire, saying, "I am the artist!" Sometimes she would wander through Dublin for days and sleep on the street.

James Joyce later commented on Nora's eccentricity, "Whatever spark or gift I possess has been transmitted to Lucia and it has kindled a fire in her brain." She was hospitalized in the 1930s for her mental illness. Her horrifying "treatments" included injections with seawater and animal serum, as well as solitary confinement.

Many eminent persons have had children with serious mental problems and have had at least one child take his or her life. Robert Frost's daughter was committed to the state mental hospital and another daughter had a "nervous breakdown." One of Albert Einstein's children was diagnosed as schizophrenic. Ambrose Bierce's oldest son committed suicide, and his other died of alcoholism at age 27. Thomas Edison had two children who became alcoholics, one of whom committed suicide. Alfred Stieglitz's daughter was psychotic and committed to a mental institution. James Joyce had two children. His son became an alcoholic; his daughter went mad and, as discussed, was admitted to an asylum for schizophrenia. Numerous other examples demonstrate the frequent problems of geniuses' children. Many of these children tried unsuccessfully to pursue careers similar to their eminent parents, but it is not clear if this played any role in their mental problems.

Recent research suggests that schizophrenia has not been eradicated from our genes because people with mild symptoms better engage both sides of their brains and use more of their brains, allowing them to excel in creative pursuits. According to Shelley Carson at Harvard University's Psychology Department, there is "no doubt that genes for schizophrenia are associated with the genes for creativity." Einstein's second son, Eduard, was a brilliant child, with very strong talents for music and languages. Alas, he suffered a schizophrenic breakdown at age 20 and was frequently institutionalized until his death in 1965. Eduard's schizophrenia gave Albert Einstein great anguish.

Carol Loeb Shloss speculates in *Lucia Joyce: To Dance in the Wake* that whatever condition Lucia Joyce had, it was worsened by family members who forced her to give up her career in modern dance— something at which she excelled. Alas, Lucia was frequently abandoned by men she loved. Her mental health declined. Lucia's brother had her committed to a hospital and insisted that she remain locked up in institutions where she was used as a human guinea pig by psychiatrists testing their nutty theories. When she was 28 years old, the Joyces put her in an asylum near Paris, and she never lived on the outside again. James Joyce loved her dearly and never believed that she was insane. He tried desperately to get her out of occupied France. Unfortunately, he died suddenly in 1941, and Lucia was abandoned to remain in mental hospitals for the rest of her life. She died in 1982 at the age of 75.

Who Really Was "Agent 488?"

The famous psychiatrist Carl Gustav Jung (1875–1961) was one of many doctors who attempted to treat Lucia Joyce. Lucia despised him, saying, "To think that such a big, fat materialistic Swiss man should try to get hold of my soul."

Jung's most famous concept, the collective unconscious, had a significant influence on psychology, philosophy, and the arts. In brief, the collective unconscious is part of a person's unconscious said to be shared between all human beings. The collective unconscious includes archetypes and symbols common to people across cultures. Jung developed this idea after studying different cultures and noting striking similarities in their patterns, myths, fairy tales, images, and religions. Traveling far afield from traditional science, Jung believed in astrology, spiritualism, telepathy, telekinesis, clairvoyance, ESP and a number of other occult and paranormal phenomena.

In Deirdre Bair's recent *Jung: A Biography*, we learn that Jung was "Agent 488," secretly working for the Office of Strategic Services, the predecessor to our CIA. In 1943, during World War II, Jung's job was to analyze the psychology of Nazi leaders for spy-recruiter Allen Dulles. In 1945, General Dwight Eisenhower, the supreme allied commander, actually read Jung's ideas for persuading the German public to accept defeat. Dulles thoroughly relied on Jung's psychological advice, including Jung's prediction that Hitler would kill himself. Later, Dulles said that "nobody will probably ever know how much Professor Jung contributed to the Allied cause during the war…[and that his work needed to remain] highly classified for the indefinite future."

Jung had an unusual start in life. His mother may have been schizophrenic. She taught him a prayer involving a winged Jesus that would take his soul, which caused the youthful Jung to be afraid of Jesus. As a boy, he experienced two personalities, which he referred to as "No. 1" and "No. 2." As a father, Jung played sadistic pranks on his children, exploding a firecracker between his daughter's legs, rendering her deaf in one ear. His daughters later said they had visions, and Jung himself heard voices and entered trancelike states while he built small-scale replicas of entire villages.

Walter Benjamin and the Arcades Project

Proust impresses me for his deep insights into the smallest items of daily life and our most subtle of emotions. German Jewish philosopher Walter Benjamin (1892–1940) was similar to Proust in some ways. His mind seemed as if it were about to explode as he pondered all aspects of life simultaneously: urban exploration, street signs, prostitution, apartment interiors, psychoanalysis, catacombs, boredom, shopping malls, Walt Disney, railway stations, Baudelaire's poetry, strange realities, and language. Benjamin assisted in the translation of Proust's *In Search of Lost Time* so that it could reach a broader audience and then spent 13 years taking notes for the "Arcades Project," 1,200 pages of insight that he called "an experiment in the technique of awakening." Then he killed himself. For Benjamin, the act of thinking opened portals to other realities. In the 1920s and 1930s, he also smoked hashish and experimented with mescaline.

We can imagine the Arcades Project as a search for truth among the hidden alleys of a large city. During Benjamin's time, arcades were glassroofed passages through blocks of buildings, lined with shops, signs, and window displays. According to Benjamin, "Arcades are houses or passages having no outside—like the dream." He traversed the arcade while his unfettered mind considered lighting design, fashion, railways, winter gardens, Saturn's rings, the relationship between art and technology, conspiracies, consumerism, feather dusters, mannequins, aquaria, wax museums, alternative histories, casinos, *Jugendstil*, gateways, hashish, Marcel Proust, and the depths of the human brain. As he wandered, Benjamin looked for shards of reality that protruded like sparkling crystals from the dark asphalt or cobblestones. He tells us again and again that the arcades are fluid places where everything strikes us "like realities in a dream"—entire mental meccas that constantly change like bits of glass in a kaleidoscope.

Probably the best book about The Project is *The Dialectics of Seeing: Walter Benjamin and the Arcades Project* by Susan Buck-Morss. She writes, "Benjamin took seriously the debris of mass culture as the source of philosophical truth....The surrealists recognized reality as a dream; *The Arcades Project* was to evoke history in order to awaken its readers from it." Dip into *The Dialectics of Seeing*, and see visions.

People attempting to read *The Arcades Project* have nearly gone mad in their quests, which can take several years. Before he killed himself, Benjamin destroyed his copy of the manuscript, and for decades the work was believed lost. Luckily, scholars found a copy, and today the work is available at Amazon.com.

About the *The Arcades Project*, *New York Times* architecture critic Herbert Muschamp wrote:

> Some of us don't read fiction [and] we will be feasting on Walter Benjamin's *Arcades Project* for years to come...Those who fall under Benjamin's spell may find themselves less willing to suspend their disbelief in fiction. The city will offer sufficient fantasy to meet most needs....[*The Arcades Project*] is a towering literary event. A sprawling, fragmented meditation on the ethos of 19th-century Paris...*The Arcades Project* captures the relationship between a writer and a city in a form as richly developed as those presented in the great cosmopolitan novels of Proust, Joyce, Musil and Isherwood.[13]

Here are some favorite Benjamin quotes:[14]

- ◆ All religions have honored the beggar. For he proves that in a matter at the same time as prosaic and holy, banal and regenerative as the giving of alms, intellect and morality, consistency and principles are miserably inadequate.

- Counsel woven into the fabric of real life is wisdom.
- Each morning the day lies like a fresh shirt on our bed; this incomparably fine, incomparably tightly woven tissue of pure prediction fits us perfectly. The happiness of the next twenty-four hours depends on our ability, on waking, to pick it up.
- Gifts must affect the receiver to the point of shock.
- Memory is not an instrument for exploring the past, but its theater. It is the medium of past experience, as the ground is the medium in which dead cities lie interred.
- The camera introduces us to unconscious optics, as does psychoanalysis to unconscious impulses.
- The true picture of the past flits by. The past can be seized only as an image, which flashes up at the instant when it can be recognized and is never seen again.
- These are days when no one should rely unduly on his "competence." Strength lies in improvisation. All the decisive blows are struck left-handed.

Benjamin believed that people became writers because they are unable to find a book already published with which they are completely happy. I think that is why I wrote my Neoreality science-fiction series that deals with fragmented realities and parallel universes—I could not find a book quite like them to read!

Sometimes I imagine myself stuck in a room with Proust's volumes at one end and *The Arcades Project* at the other. Perhaps I could be happy for the rest of my life in this self-imposed exile. Could you? For Benjamin, the enclosed shopping malls of the 19th century were a material replica of a human collective unconscious. Contemplated along with Proust's novels of involuntary memory, we'd enter wonderful, quaquaversal realms of thought and imagination.

Psychedelic Shakespeare and DMT Elves

Far from the urban landscapes of Benjamin lie the castles and deep forests of William Shakespeare. In Shakespeare's works, characters often exist at the intersection of our universe and the world beyond, and the characters are never quite sure what is real and what is imaginary. As just one example, when Macbeth sees a ghostly dagger floating above him, he ponders, "Art thou not, fatal vision, sensible to feeling as to sight, or art thou but a dagger of the mind, a false creation, proceeding from the heat-oppressed brain?"[15] Later Macbeth sees ghosts and phantom bloody children.[16]

Yes, in psychedelic Shakespeare, our mental lives continually merge with the external world. Life is a dream. Characters walk around as if in a trance or in a drug-induced stupor. DMT-like fairies and nymphs are everywhere. In *A Midsummer Night's Dream*, his elves are almost as small as insects. In other plays, Shakespeare tells us that elves are little people. Shakespeare's supernatural subjects—the elves, the weird sisters in *Macbeth*, the ghost in *Hamlet*, Prospero's magical spirits—all conjure images of another realm so close to ours that we are able to touch it in our daily lives.

Why have elves permeated the minds of men since the dawn of time? Why are elves seen under the influence of DMT (Chapter 4) and by people afflicted with Charles Bonnet syndrome (Chapter 5)? Perhaps we will never know. Elves were pervasive in Germanic mythology and imagined as small, attractive people living in forests. The Norse myths referred to light-elves (Liosálfar), dark-elves (Döckálfar), and black-elves (Svartalfar) who were roughly of human size. The tomtar, vittror, and älvor are slightly more modern Scandinavian elves. The älvor dance in the meadows at night and leave behind traces in the form of älvringar (elf circles).

In Chapter 4, we mentioned that the DMTverse and DMT machine elves seemed to operate on a different time than ours, or at least many psychonauts reported that that the DMTverse seems to exist independent of our ordinary flow of time. Similarly, legends tell us that if a human watched the dance of the älvor, a few minutes might pass in the dance while years flowed by in the rest of the real world.

Tolkien also tells of time distortion in the elven forest of Lothlorien where the elves never grow old. According to the *Lord of the Rings,* Lothlorien trees' leaves turn to gold just as in the DMT world, which is frequently populated by objects of gold. "The boughs are laden with yellow flowers," Tolkien writes, "and the floor of the wood is golden, and golden is the roof, and its pillars are of silver..." Lothlorien is a very DMT-like environment!

A Midsummer Night's Dream has had a rapid and enduring effect on human concept of elves ever since its publication, and has inspired such authors as Ben Jonson (1572–1637) and Milton (1608–1674). Long before Shakespeare, Chaucer (1343–1400) said that fairies came into being "far backward of time," and they faded away with the flourishing of

Christianity. Now, elves and spirits are rarely seen in our everyday lives, but still seem to inhabit our deepest dreams:

> In Wonderland they lie,
> Dreaming as the days go by,
> Dreaming as the summers die:
> Ever drifting down the stream—
> Lingering in the golden gleam—
> Life, what is it but a dream?[17]

The Chanting of the Sausalito Mystery Fish

One spring day, after my kung fu and tai chi class ended, several classmates visited a local Buddhist temple. I had been studying these martial arts for about five years, but I wanted to better understand some of the more spiritual aspects. When we entered the shrine, I noticed a fish-shaped object on the side of the altar. I was told that the wooden fish is an important Buddhist instrument for meditative practice. Legend says that fish do not close their eyes and thus are attentive throughout night and day. Buddhists carve beautiful fish and tap on them during chanting to remind us not to doze off or lose awareness. Later that day, I also noticed some wooden fish suspended outside the dining hall and in other areas.

Alas, I would not make a good Buddhist because my mind continually drifts. While I sat on a prayer mat, I began pondering a recent mystery involving the chanting of the Sausalito mystery fish.

Scientists and homeowners living on the California coast had been hearing an eerie sound from the waters—a perfect A-flat that droned on for hours, resembling the chanting of Buddhist Lamas or perhaps something more sinister. In fact, this deep, low-pitched sound seemed like something right out of a Stephen King movie or The X-Files. In the 1980s, some conspiracy theorists blamed the ghostly drones on secret experiments performed by the Army Corps of Engineers. Others said the sound came from alien life forms. When I had first heard the mysterious drone, I immediately thought of the Acoemeti, a group of monks who provided nonstop choral singing in the 5th century. Their never-interrupted relays could go on for days as they affirmed their conviction that God should be perpetually praised.

We now know that the humming male toadfish, also known as the midshipman, produces the loud and incessant underwater droning that was a mystery for years. I've seen the photos. The foot-long fish with its flat head is quite ugly looking. However, the sides and belly of the fish have patches of bioluminescence that are pretty in the dark. The colorful patches might have evolved to help the toadfish attract food, like small fish and crustaceans. The loud drone is actually the sex call of the fish, and is now leading scientists to wonder just how the fish can produce a loud sound for so long without tiring. Luciano Pavarotti would be proud.

The toadfish's extraordinary muscular strength above its swim bladder allows it to drone on seemingly forever, its muscles vibrating around 6,000 times a minute. Researcher Kuan Wang of the National Institute of Arthritis, Musculoskeletal, and Skin Diseases believes that by studying the muscles of the toadfish, he may find ways to treat human muscle-weakening genetic diseases like nemaline myopathy.

Despite our nascent understanding, mysteries of the toadfish still remain. No one knows for sure why the fish developed such a loud love call—however, we can make some guesses. The toadfish comes in three "genders": female, Type 1 male, and Type 2 male. Type 1 males are dominant and sing the loudest and longest. Type 2 males are quieter. The Type 1 male is more likely to attract a gal to his rock nest, in which she deposits eggs. (Incidentally, the male toadfish also uses its mating call to lure females to nests in discarded coffee cans.)

However, the loud call of Type 1 males does have some disadvantages. Type 2 males, which mimic the females in appearance, may attract a Type 1 male, which promptly begins to copulate with the Type 2, at which time the Type 2 gets a chance to fertilize the eggs in the nest. Thus, by pretending to be a gal, the meek males can sometimes fertilize eggs along with the Type 1 Schwarzenegger. Note that the Type 2 males, although smaller, have huge reproductive organs relative to Type 1 fish. Imagine a midget with a schlong the size of the Graf Zeppelin.

Another problem with the loud love song (Type 1) is that it not only attracts the female fish, but also sharks. I wonder what behavior of humans most closely corresponds to the loud drone of the toadfish, with its charms and perils. Incidentally, the toadfish is one of many fish that use flamboyant methods for attracting mates. The male swordtail characin, for example, has long stringy filaments that dangle from its gills. These tendrils resemble the daphnia worm, which the characin love—and the male displays the fake worms to lure females with which he can mate. This leads me to the vast panoply of methods humans have evolved to attract mates, which we discuss in the next section.

Attracting Women

If the singing toadfish can find mates, why do some men and women find it so difficult? I am always curious about what makes people attractive to one another, and, in particular, what attracts women to certain men. Obviously, many women are attracted to confident men and men of accomplishment. But the attraction equation is so much more complex than this.

Let's discuss actual "recipes" people can use to become attractive to potential mates. As with so many topics, Marcel Proust was an expert at this, at least from a theoretical standpoint. He wrote, "There is no doubt that a person's charms are less frequently a cause of love than a remark such as: 'No, this evening I shan't be free.'" In other words, above all else, never appear to be needy. Telling someone you are too busy to see them is the ultimate aphrodisiac.

Prostitutes hold an interesting position for Proust. They want to attract a man but do not have the benefit of being able to say they "aren't free." They must be available, and are thus never as desirable as a woman who can tell a man that she is busy. Proust writes, "If prostitutes...attract us so little, it is not because they are less beautiful than other women, but because they are ready and waiting: because they already offer us precisely what we seek to attain."[18] Of course, Proust may have wished to ponder why, if prostitutes "attract us so little," they have proliferated so much. Perhaps he refers to love and not lust.

An even better recipe for mate acquisition is one that appeared in a 2004 *New York Times* article "He Aims! He Shoots! Yes!!"[19] This particular recipe has been demonstrated to be essentially foolproof for men who want to attract women in social settings and, in particular, to coax even the most attractive women to go out on a date with the man. I realize that this recipe sounds a little contrived or demeaning, yet we can all be enthralled by its scientifically proven efficacy. The algorithm has three steps:

> *Step 1.* If a man wants to meet a woman in a group of both men and women, the man must approach the group and befriend the *men* first. Do not engage the woman. Do not glance at her. (Example: you can chat with the guys and show them something interesting like a magic trick, without directing your attention to the woman.)

Step 2. Use one "neg" on the woman. A "neg" is the invention of Erik von Markovik, developer of this recipe. As the *New York Times* reports, "Neither a compliment nor an insult, a neg holds two purposes: to momentarily lower the woman's self-esteem and to suggest an intriguing disinterest."[20] Negging should primarily be used for "exceptionally beautiful women used to a steady stream of compliments." An example neg is: "Nice eyelashes. Are they real?"

Step 3. Eventually, the woman will clamor for the man's attention. Engage the woman, but you should never introduce yourself *first* to the woman. You must wait for the woman to introduce herself, which is one of several possible "indicators of interest" or IOIs. If the man receives three IOI's, then the option to see the woman becomes possible.

Studies have shown that men who use these tough rules have an extremely high success rate of attracting women, even if the man is not handsome. Isn't this amazing? Even more interesting are the dozen women who work for the New York–based Web site Wingwomen.com. They earn up to $30 an hour to accompany single men to bars and help them meet other women. These "wingwomen" have a very high success rate in ensuring that the men end up with the phone numbers of attractive woman.

According to Web-site founder and computer programer Shane Forbes, women like to compete and are more attracted to men "who have other women surrounding them because they want what they can't have." Additionally, Forbes has found that women tend to lower their defenses around men who have other women around them. Women tend to see these men, accompanied by wingwomen, as having a "seal of approval" and being nonthreatening.

Proust saw the value of jealousy as a tool for reviving a relationship that was getting boring. He wrote, "When you come to live with a woman, you will soon cease to see anything of what made you love her; though it is true that the two sundered elements can be reunited by jealousy."

Word Salad and the Future of Computer Spam

Today, when I opened my computer mail Inbox at work, I had dozens of unsolicited electronic messages, known as "spam." Most people who use a computer are familiar with these bothersome advertisements.

Spam sending is easy to automate: computer programs can send out millions of e-mail messages in hours with little cost to the sender.

As the 21st century progresses, the bulk of e-mails will be meaningless. Already, many subject lines convey no meaning at all in order to avoid and trick spam filters that are designed to eliminate traditional spam. Thus, crazy, unpredictable subject lines like *"visual, rhododendron, purveyor, noodle"* are becoming common, although I wonder how many people really open these mails. If you do open them, the mails get even harder to decipher, and we see:

Gkmlwrtw We offer VI@GRA at 80% D1sc0unt! m lcgpggbksr
or
Cheap Meds Viagr@,Vali(u)m, X(a)n@x, Som@ Di3t Pills

Again, the purpose for obscuring the spelling of words is to confuse the spam-stopping software that might ordinarily discard a clear message that read: *We Offer Viagra at 80% Discount!* What I find confusing is that it is becoming impossible for human beings to read the e-mails, so what's the point of sending them?

When I search with Google, I find thousands of instances of obfuscated spellings that have metastasized into Web pages and bulletin boards like: *vi@gra, v-i-@-g-r-a,* and *viagr@.* I recently received a mail with the message *"Free adult epikaihlklzya,"* which did not excite me in the least.

Here's a subject line I got today, and the product the e-mail sells:

Subject: pitman defuse hellenic bellwether
    Pénis Enlargemént
(Notice the modified "e" to escape spam detectors.)

Here are some more:

Subject: adultery compagnie dichotomize farthest
    Idéntity Théft Prótector
Subject: calorimeter fishmonger foppish
    Ciális (something even better than Viagrá)
Subject: excoriate bothersome plywood
    Enháncemént Oil - Gét stiff as a róck in 60 séconds!

This insertion of random text in e-mail spam is known as "word salad," which usually consists of random words from a dictionary or text fragments from a work like *Hamlet*. Incidentally, "word salad" is a phrase first used in psychiatry to describe the seemingly random use of words by certain schizophrenics. For example, the question "Why do people believe in an afterlife?" might elicit an unintelligible response like "Because drink a star in death, my mind is dripping—help me amber mantis. Aren't spinal cords brave? I love protons, now." Perhaps someday neuropsychologists will be able to partially decode such utterances.

The better the spam filters get, the more cryptic the message become. As sad as it is to find these indecipherable misfits in our mailboxes, at least we can be happy that the spammer's messages are becoming so difficult to interpret, that in a few years they will not be readable, if current trends continue.

I have a spam mystery question that maybe one of you can solve. I read in the March 13, 2004, *New Scientist* about a man who tried to order every product that was sent to him in the form of 200 spams. None of the offers he received turned out to be real. He said that none of the advertisements carried Web, e-mail, or postal addresses that worked. None had a mechanism to relieve him of his money. None were actually trying to sell anything. So, what exactly is the purpose of this spam? Who benefits from the communications that serve no commercial purpose? Were the messages he received simply the results of people "testing" spam-sending software? Could they be decoys to slow down law enforcement in their search for actual scams?

My colleague Hannah M. G. Shapero notes that spam is starting to intrigue artists and writers. She notes, "Spam is, in its own weird way, a kind of multi-universal poetry. The aggregates of unusual or random words and strange languages speak to us like a kind of globalized static, with subject lines often appearing in transliterated Russian, Polish, and Latin."[21]

Hannah ponders the evolution of spam over the next thousand years:

Much spam comes from fake "personal names,"constructed by name-generators that can create as many identities as there are names to combine, even selecting for ethnicity so the recipient can receive Latin, Anglo, or Asian names. Consider how many new "friends" you have, virtual crowds of billions sending you personal mail. Every day, trillions of new human identities are being created and named, just to send you spam.

These days, names are also assembled from the pool of words contained in the bodies of the e-mails. So I get mail from astonishing and hilarious personalities like "Firmament V. Explosive," "Obtainable P. Schizoids," or "Cocktail A. Beatitude." Hi there, Cocktail! How's Mrs. Beatitude?

Someone should preserve an archive of spam for ages to come. A thousand years from now, if civilization still exists and is capable of reading the records on today's ancient media, people will wonder about our society and what these messages say about us. It will appear that our males are plagued with impotence and small genital size. Our females are sexually unresponsive until they encounter large male genitals.... We are all eager to invest in mortgages. There is an endless supply of money in Africa. And many of us chose to write in incomprehensible language clusters, which still elude meaning even though all the words can be found in dictionaries.[22]

I leave you with a spam I received with the subject line "Supply of Cow Medicine." I wonder how many people are interested in investing in cow medicine from the Ivory Coast?

ADVERTISING, MARKETING AND INTERNATIONAL
COMMUNICATIONS SERVICES (AMICS)
Office: Rue Paris Villa koumassi 02 BP 2563.
0022507060855
Subject: SUPPLY OF COW MEDICINE
Dear Sir,
    I came in contact with your esteem company through your country's Chamber of Commerce. I became interested and decided to write you and tender a request for detailed information about your company. I have a business of cow medicine supply with a desire to have a link-up for Trade relationship with you for mutual benefits. I am Mr. JAMES the corporate affairs Public Relations Manager of an Ivory Coast based public Relations Firm-ADVERTISING, MARKETING AND INTERNATIONAL COMMUNICATIONS SERVICES (AMICS). During an international business seminar which was held in Quebec Canada between the 7th and 19th October 1999 in which I was privileged to attend.
    I came in contact with prominent international business magnets-among whom was Alhaji SHERIF AL-MUSTAFA. The President of Mc Bella Group of Companies (S.A.R.L.) areknown Muslim Imam and a successful businessman, with Cattle Farms in Belgium, Haiti, Middle East and North African sub-Regions. Alhaji Sheriff Al-Mustafa is at the moment the greatest supplier of Cattle to the whole of middle East bloc. After a bit introduction of my self and profession, Alhaji Sheriff Al-Mustafa was delighted to come in contact with me

and with a profound interest to do business with me in all capacity. Though Alhadji Sheriff Al-Mustafa is an accomplished businessman, but complained bitterly on the huge amount of money he spends on the purchase of a particular cow medicine, but very vital COW medicine which according to him costs him as much as $5,000 US dollars per carton with a minimum Order of 1,500 cartons monthly. Alhadji Sheriff Al-Mustafa was exceedingly very much delighted when I told him that my company can source for the supply of this medicine at very reasonable and unique prices.

Alhaji Sheriff Al-Mustafa has suspended further contacts with my boss. But now the question is this: (1) Can you handle this project? (2) What will be my commission? If the answer to the above questions is affirmative-do not hesitate to contact me immediately to further discussion on the project. Then you can contact me on the above lines.

Awaiting your urgent response for us to proceed.

Regards,

Mr. James

Dropping Pennies from the Empire State Building

Readers frequently write to me, asking all kinds of questions, ranging from mathematics to religion. Sometimes I answer the questions directly. At other times, if my schedule is too busy, I refer fans to my Internet Think Tank, which has over 600 members. Almost any question can be answered in my Think Tank, where we collectively function as a hive mind to solve problems.

One question I'm repeatedly asked deals with death, New York City, and pennies. Just the other day, "Teja" from Slovenia asked me: "If I accidentally drop a penny from the Empire State Building, would it kill someone walking on the street below? Would it literally go through the skull and brain and out the bottom of the chin? My friend Nina accidentally dropped a penny and has been worried ever since."

Teja, you can rest easy. The penny falls about 500 feet before reaching its maximum velocity: 57 miles an hour. A bullet travels at ten times this speed. The penny is not likely to kill anyone. Updrafts also tend to slow the penny. Also note that the penny is not shaped like a bullet, and the odds are that the tumbling coin would barely break the skin.

Exploding Closing Credits in Movies

Most of you have probably seen closing credits on TV or in the movies. They come at the end of a show and list all the cast and crew involved in the production. Have you ever noticed that the number of names on film closing credits is growing tremendously?[23] You frequently see the names of gaffers, grips, and best boys listed on closing credits, in addition to the caterers, minibus drivers, and even assistant hairdressers.

I have finally figured out what a "gaffer" is—he or she is an electrician responsible for lighting on a movie or TV set. The "best boy" is the chief assistant to the gaffer. "Grips" are crew members who move the equipment and assemble it on the set or on location.

I've made a list of some of the most unusual credits in movies that I've seen: contact lens consultants (*Wolf*), equine makeup (*Black Beauty*), standby flute advisor (*Hilary and Jackie*), crow wrangler (*The Crow*), physiotherapist (*Alien 3*), and plaster foreman (*Alien Resurrection*).

Randy Kennedy wrote an article for *The New York Times* that expressed amazement at the length of the closing credits in *The Lord of the Rings: The Return of the King*.[24] Intrigued, I decided to stay to the very, very end of *The Return of the King*, and found that I sat for over nine minutes during the credits. These days, you'll see diverse people ranging from cast aroma therapists to food stylists to assistant accountants. I predict that in a few years, we will be seeing the names of the boyfriends of the manicurists for the stunt double's dentist. At the current rate of growth, in the year 2035, the duration of the closing credits will be equal to that of the actual movie.

Here are some of my favorite credits in *Return of the King*: Let us hereby give thanks to the Accountants and their Assistants, the Hammerhands, the Standby Greens, Apprentice Assistant Editors, Human Resources Assistant, the Focus Pullers, the Clapper Loaders, the Trainee, the Caterers, and consultant for Cultural Fighting Styles.

Sometimes when I sit through the closing credits, watching as all the actor names scroll by, I feel as if I am reliving parts of the movie, as if I'm experiencing a "life review"—the flash of memories that people experience when having near-death experiences. People pulled back from the brink of death sometimes tell of a sequence of images in which they see and re-experience major and trivial events of their lives. Closing credits make me think of the ending of someone's life, funeral music, and eulogies. These cheerful thoughts bring us to the next section.

Near-Death Experiences, Proust, and the Dissolution of Self

"Near-death experiences" (NDEs) are sometimes reported by people who have nearly died or have been clinically dead and revived. During the experience, people sometimes have the sensation of floating above their bodies and seeing a tunnel of light.

People having near-death experiences sometimes say that they see a silver cord connecting their bodies to "heaven" or to their "souls." They believe that the cord is similar in nature to a newborn's umbilical cord and point to the existence of the silver cord in the Bible: "Remember him, before the silver cord is severed…and the dust returns to the ground it came from, and the spirit returns to God who gave it."[25] According to the Jewish Talmud, the soul is usually severed from the body immediately after death. If the person has led a spiritual life, this separation is effortless, "like removing a hair from milk" (Talmud, Moed Katan, Chapter 3). On the other hand, sinful people have painful separations and sometimes are subject to the tormenting experience of watching their bodies decay before a clean break is made.

Kevin Williams, creator of the extensive Web site near-death.com, has catalogued dozens of sightings of silver cords at his site. Respondents said, "I saw a long silver cord coming out of my spirit body, right through the fabric I was wearing," or "Almost instantly I felt re-entry into my body through the silver cord at the top of my head," or "At that moment I noticed a silver cord, attached around the navel," as well as many more such reports. The image of a silver, twisting cord is echoed in the movie *Donnie Darko*, which inexplicably shows a liquidlike tube protruding from peoples' bellies and pointing in the direction that person will move in the near future. Donnie Darko sees his own lifeline stretching from his belly, as if his actions have been predetermined, and he's a pawn, trapped in the jejune jardinière of time.

Similarly, Marcel Proust in his epic *In Search of Lost Time* describes "a rope let down from heaven to draw me up out of the abyss of non-being…" Let's discuss the amazing Proust in the next section.

My Obsession with Marcel Proust

As I mentioned in other chapters, Proust was a fascinating writer, best known for his seven-volume work, *In Search of Lost Time* (older English translations have used the title *Remembrance of Things Past*). The novel draws heavily on the Proust's own life and experiences, but he made sure that the main character also had experiences that Proust never had. Most of the book's characters were metaphors for actual people he had encountered in real life, and the characters often combine traits of several different friends or lovers. The closeness of the novel to his life is revealed on two occasions near the end of the novel, when he mentions that "Marcel" is the first name of the narrator.

Proust never actually liked the original title *Remembrance of Things Past*, because he did not think it was an accurate translation of his original French title. The phrase *"remembrance of things past"* comes from a Shakespearean sonnet, "When to the sessions of sweet silent thought / I summon up remembrance of things past…"

In Proust's book, the novel's protagonist, often known simply as the Narrator, gradually realizes that people go through life living on the surface, busily pursuing materialistic goals that make them unhappy. Proust realizes that he can convert his own memories into art so that his readers can learn by, and resonate with, his experiences.

Amazingly, his grand work spans over 3,200 pages and features more than 2,000 characters. About a third of the book involves the narrator attending dinners, receptions, or parties!

According to many scholars, Proust was the greatest novelist of the 20th century, and *In Search of Lost Time* is the greatest fiction ever written. Even though the book sometimes seems to meander or get stuck in little pocket universes filled with elegant dinner parties, Proust insisted that he wrote the last pages of the novel immediately after the first page. He had a plan, a goal, a vision, a sense of control. Although the task was daunting, the master writer was in charge and fully in command.

British novelist Virginia Woolf (1882–1941) became so enraptured by *In Search of Lost Time* that she wrote, "Proust so titillates my own desire for expression that I can hardly set out the sentence. Oh, if I could write like this!" To Woolf, Proust's work was the greatest of adventures. She felt that nothing could be written after Proust. "How has someone solidified what has always escaped—and made it too into this beautiful and perfectly enduring substance? One has to put the book down and gasp."

Proust's mom was Jewish, his dad a Roman Catholic physician. His younger brother was a surgeon and author of the world-renowned *The Surgery of the Female Genitalia*. Proust himself was often ill, suffering from asthma and other ailments. Despite his frailty, in his younger days, he was quite social and well integrated into the "high society" of the time, and he describes this social world in his novel. However, during his thirties, he began to withdraw from society, and devote himself entirely to his writing.

By about 1910, Proust essentially lived by night and slept by day, rarely venturing from the bed in his cork-lined room. From this time to his death in 1922, his novel was the most important aspect of his life. While in bed, he sent out periodic reports on the number of pages he had reached. In fact, Proust spent years lying in his narrow bed, writing this long novel without even an adequate bedside lamp. He believed that we only really think when distressed. Thus, we shouldn't search for happiness but rather pursue ways to be "properly and productively unhappy."[26] Each problem contains a gift for us, something to solve. Proust also wrote:

- ♦ "If a little dreaming is dangerous, the cure for it is not to dream less but to dream more, to dream all the time."
- ♦ "The real voyage of discovery consists, not in seeking new landscapes, but in having new eyes."
- ♦ "All our final decisions are made in a state of mind that is not going to last."

Proust always believed that we will encounter certain personality types repeatedly during the courses of our lives, and that this is actually a positive feature of the cosmos: "Aesthetically, the number of types of humanity is so restricted that we must constantly, wherever we may be, have the pleasure of seeing people we know...." (I wonder if DMT psychonauts will ever get to the point where they are continually reunited with the same machine-elf personalities to which they can form happy or productive bonds.)

In Search of Lost Time has been translated into more than 35 languages and has never been out of print since its initial publication. The work's amazing popularity is due to Proust's psychological insights into how humans view the deepest essence of reality.

Proust's sentences are often quite long, and I believe the longest occurs in the fifth volume of *In Search of Lost Time*. If arranged in a single line of standard-sized text, the text line would stretch for more than 13 feet! It begins, "A sofa that had risen up from dreamland between a pair of new and thoroughly substantial armchairs...." Another impressive sentence is 958 words long and starts, "Their honor precarious, their liberty provisional, lasting only until the discovery of their crime; their position unstable..."

Alfred Humblot, who was head of one potential publisher for Proust, rejected *In Search of Lost Time*, saying "I may be dense, but I fail to see why a chap needs thirty pages to describe how he tosses and turns in bed before falling asleep." Jacques Madeleine, another reviewer, said, "At the end of 712 pages of the manuscript...one doesn't have a single clue of what this is about." Yet another said to Proust:

I only troubled myself so far as to open one of the notebooks of your manuscripts; I opened it at random, and as ill luck would have it, my

attention soon plunged into the cup of camomile tea on page 62—
then tripped, at page 64, on the phrase…where you speak of the
"visible vertebra of a forehead."[27]

Eventually, Proust was forced to pay for the publication *In Search of
Lost Time* himself.

Because Proust's asthma seemed to be worst during the day, he began going to sleep at seven in the morning and waking up at four in the
afternoon. As described in Alain de Botton's *How Proust Can Change
Your Life*, Proust's windows and curtains were always shut. Eventually,
he never saw the sun, breathed fresh air, or exercised. He ate only one
meal a day—a huge one consisting of eggs in cream sauce, a chicken
wing, multiple croissants, and a bowl of French fries. He used twenty
towels a day to wash himself.

Living in the π Matrix

I recently asked members of my Pickover Think Tank to consider a
concept that had been on my mind for many years. My contention was
that somewhere inside the digits of pi is a representation for all of us—
the atomic coordinates of all our atoms, our genetic code, all our thoughts,
all our memories. Given this fact, all of us are alive and happy, in pi. Pi
makes us live forever. We all lead virtual lives in pi. We are immortal.

This means that we exist in pi, as if in a matrix. Thus, romance is
never dead. Somewhere you are running through fields of wheat, holding hands with someone you love, as the sun sets—all in the digits of pi.

Pi, symbolized by the Greek letter π, is a "transcendental number." It
is a never-ending, patternless sequence of digits. Each digit appears with
equal frequency. Here are the first few digits:
π = 3.1415926535 8979323846 2643383279 5028841971 6939937510
    5820974944 5923078164 0628620899 8628034825 3421170679
        8214808651 3282306647 0938446095 5058223172

Is it really true that we all live happy lives, coded in the endless digits
of π? I believe so, although many people have debated me on this subject. Recall that the digits of π (in any base) not only go on forever but
seem to behave statistically like a sequence of uniform random numbers.
In short, *if* the digits of π are normally distributed, somewhere inside π's
string of digits is a very close representation for all of us.

You can read a long group discussion of this topic, which I initiated at my Web site www.pickover.com/pi.html. Here we debate all my assumptions about how pi lets us transcend our Earthly existences. Incidentally, the first person to uncover an infinite product formula for pi was French mathematician François Viete (1540–1603). This remarkable gem of an equation involves just 2 numbers—π and 2:

$$\pi = 2 \cdot \frac{2}{\sqrt{2}} \cdot \frac{2}{\sqrt{2+\sqrt{2}}} \cdot \frac{2}{\sqrt{2+\sqrt{2+\sqrt{2}}}} \cdots$$

My readers often ask me if it is true that we can find consecutive digits, like $1, 2, 3, \ldots 1,000,000$, all neatly in a row in the decimal digits of π. Certainly this is true if we assume modern mathematical conjectures are correct. We can even search for some of the first few consecutive runs using computer searches available on the Web. The string 123 is found at position 1924 counting from the first digit after the decimal point. The "3." is not counted. 1234 is found at position 13,807. 12345 is found at position 49,702, and so forth. You can do further searches of this kind at Dave Anderson's π Web site: http://www.angio.net/pi/piquery.

Quantum Immortality and Parallel Universes

If you are not comfortable knowing that you are immortal in pi, maybe you can be comfortable with the idea of "Quantum Immortality." To understand this concept, first we must understand the many-worlds interpretation of quantum mechanics. Hugh Everett III's doctoral thesis, "Relative State Formulation of Quantum Mechanics" (reprinted in *Reviews of Modern Physics*), outlines a controversial theory in which the universe at every instant branches into countless parallel worlds. However, human consciousness works in such a way that it is only aware of one universe at a time.

This theory is called the many-worlds interpretation of quantum mechanics and holds that whenever the universe ("world") is confronted by a choice of paths at the quantum level, it actually follows both possibilities, splitting into two universes. These universes are often described as "parallel worlds" although, mathematically speaking, they are orthogonal or at right angles to each other. In the many-worlds theory, there are an infinite number of universes, and if true, then all kinds of strange worlds exist. In some, Hitler won World War II or the Russians put a man on the moon before the Americans. Sometimes, the term "multiverse" is used to suggest the idea that the universe that we can readily observe is only part of reality that comprises the multiverse, the set of possible universes.

Aside from my own novels in the "Neoreality series" (Chapter 8), my favorite tales of parallel worlds are those of Robert Heinlein. For example, in his science-fiction novel *The Number of the Beast* there is a parallel world that appears identical to ours in every respect except that the letter "J" does not appear in the English language. Luckily, the protagonists in the book have built a device that lets them perform controlled explorations of parallel worlds from the safety of their high-tech car. In contrast, the protagonist in Heinlein's novel *Job* shifts through parallel worlds without control. Unfortunately, just as he makes some money in one America, he shifts to a slightly different America where his money is no longer valid currency, which tends to make his life miserable.

Those who believe in "Quantum Immortality" say that the many-worlds interpretation of quantum mechanics implies that a conscious being can live forever. The cancer you have will not kill you. Your fatal car accident ten years from now will never take place. The theory also means that suicide bombers continue to exist, even after their backpacks explode. The strange logic for quantum immortality becomes clear in the following paragraph.

Suppose you are on an electric chair, which your executioner turns on. In almost all parallel universes, the electric chair will kill you. However, there is a small set of alternate universes in which you somehow survive—for example there may be a power failure at the instant the executioner pulls the switch. Or the governor will grant you a pardon. The idea behind quantum immortality is that you are alive in, and thus able to experience, one of the universes in which the electric chair malfunctions or you are set free, even though these universes form a small subset of the possible universes. In this way, you would appear, from your own point of view, to be living forever. Although Quantum Immortality is highly speculative, I am not aware of any laws of physics that this violates. I should point out that physicists postulate many different kinds of parallel universes. Mathematicians dating back to Georg Bernhard Riemann (1826–1866) have studied the properties of multiply connected spaces in which different regions of space and time are spliced together. Physicists, who once considered this an intellectual exercise for armchair speculation, are now seriously studying advanced branches of mathematics to create practical models of our universe, and to better understand the possibilities of parallel worlds and travel using wormholes and by manipulating time.

Perhaps God sees all universes. The many-worlds theory suggests that a being existing outside of spacetime might see all conceivable forks, all possible spacetimes and universes, as always having existed. How could a being deal with such knowledge and not become insane? A God would see all Earths: those where no inhabitants believe in God, those where all inhabitants believe in God, and everything in between. According to the many-worlds theory, there could be universes where Jesus was son of God and universes where Jesus did not exist. Perhaps free will for God, or saying that God can make choices, merely means He shifts His consciousness from one branch of the multiverse to another whenever He makes a choice, like a ball rolling down a forked path and going right or left. Of course, how an omniscient God shifts consciousness would be difficult for us to imagine. Perhaps the universe that God does not choose simply ceases to exist.

Much of Everett's many-worlds interpretation is concerned with events on the submicroscopic level. For example, the theory predicts that every time an electron either moves or fails to move to a new energy level, a new universe is created. Currently it is not clear the degree to which quantum (submicroscopic) theories impact reality at the macroscopic, human level. Quantum theory may even clash with relativity theory, which forbids faster-than-light transfer of information. For example, quantum theory introduces an element of uncertainty into our understanding of the universe, and it states that any two particles that have once been in contact continue to influence each other no matter how far apart they move, until one of them interacts or is observed. In a strange way, this suggests that the entire universe is multiply connected by faster-than-light signals. Physicists call this type of interaction "cosmic glue." The holy grail of physics is a full understanding of the comic glue while reconciling quantum and relativistic physics.

We live in a visible universe easily encompassed by a sphere 100 billion light-years across, with a finite number of configurations for the matter and energy contained within. Let's imagine our visible universe as a gigantic bubble floating within our larger universe. (We cannot see infinitely far because the Universe has a finite age and because information cannot travel faster than the speed of light.) If our universe is infinite, as many modern physicists believe, then identical copies of our bubble must exist, with an exact copy of our Earth and of you. According to physicist Max Tegmark, on average, the nearest of these identical bubbles is about 10 to the 10^{100} meters away. Not only are there infinite copies of you, there are infinite copies of variants of you. It is almost certain that right now you have red eyes and are kissing someone who speaks Etruscan with long fangs in some other bubble. If we accept the notion of an infinite universe—which is suggested by modern theories of cosmic inflation— infinite copies of you exist, altered in fantastically beautiful and ugly ways.[28]

Some physicists have suggested that higher spatial dimensions may provide the only refuge for intelligent life when our universe eventually dies in great heat or cold. Michio Kaku, author of *Hyperspace*, suggests that "in the last seconds of the death of our universe, intelligent life may escape the collapse by fleeing into hyperspace." Our heirs, whatever or whoever they may be, will explore these new possibilities. They will explore space and time. They will seek their salvation in the higher universes. Will God be with them?

Could Jesus Multiply Two Numbers?

I have three questions to pose to you. Would Jesus of Nazareth ever have worked with a negative number, like -3? What kinds of written numbers did Jesus, or a comparable figure of his era, use? Could Jesus multiply two numbers?

Let's first tackle the question of Jesus and negative numbers. Jesus never worked with a negative number like -3. The concept of negative numbers started in the seventh century AD. At this time, we first see negative numbers used in bookkeeping in India. The earliest documented evidence of the European use of negative numbers occurs in the *Ars Magna*, published by Italian mathematician Girolamo Cardano in 1545. Al-Khwarizmi, who was born in Baghdad, discovered the rules for algebra around 800 AD. Obviously, there is quite a bit of surprisingly simple mathematics that was not around in Jesus's time.

What kinds of numbers might Jesus have used? According to my colleague and scholar Daniel Dockery, Jesus spoke Aramaic, and he expects that Jesus used the Aramaic/Hebrew number system where alphabetic characters also served as their numbers. Because some of the apocryphal and pseudepigraphic infancy gospels tell tales of Jesus having discussed the symbolism of the Greek and related alphabets, one might also argue that he could have written using the Greek number system, which likewise used its alphabet for numerical digits.

If one considers the text of the New Testament as definitive, reliable, or historical, all numbers that appear in passages with references to Jesus in the four gospels are written out in Greek (e.g., *eis/mian* [one], *duo/duos* [two], *treis/trisin* [three], *tessares* [four], *hex* [six], *hepta* [seven], *okto* [eight], *heptakis* [seven], *ennea* [nine], *deka* [ten], *eikosi pente* [twenty-five], triakonta [thirty], *hekaton* [one hundred], *hebdomekontakis heptai* [seventy times seven], *dischilioi* [two thousand], *pentakischilioi* [five thousand], etc.) Most numbers in the text of the Bible tend to be written out, though there are a few exceptions, such as the infamous 666 of the Apocalypsis, written with the three Greek letters *chi, xi*, and the antiquated *stigma*. In the Greek numeral system, the letter *chi* has a value of 600, *xi* 60, and the *stigma/digamma* a value of 6, so that the three letters appearing together as a number have the combined value of 666.

Finally, it is very likely that Jesus could multiply two numbers. In Matthew 18: 22 we find, "*legei auto ho Iesous Ou lego soi eos heptakis all'eos hebdomekontakis epta.*" Or, in Jerome's Vulgate: "*dicit illi Iesus non dico tibi usque septies sed usque septuagies septies.*" Today we translate this as "Said Jesus: To you I say not 'til seven times,' but 'until seventy times seven.'" Because both 7 and 70 can have symbolic meanings, the meaning may not be literal, but nevertheless it is an example of multiplication.

The Bible does not make it clear whether Jesus or his listeners would have been able to give the exact answer. Much earlier, in Leviticus 25: 8, we find. "Seven weeks of years shall you count—seven times seven years—so that the seven cycles amount to forty-nine years." Therefore, we know these people could do at least 7×7. However, we must not lose sight of the possibility that the Biblical translators introduced the terms.

Additionally, conversion between monetary systems like Roman sesterces, Jewish shekels, and Persian darii probably required notions of multiplication and division. It is likely that Jesus was aware of the concept of debts and interest charged on debts.

Jesus would not have used a symbol for zero, because neither the Hebrew, Aramaic, nor Greek number systems had a character to represent the number 0, as it was not required by their nonpositional number systems.

The Wishing Project

I am obsessed by people's wishes and what they teach us about different ways of looking at the world. I remember an ancient parable that describes a conversation between God and two human adversaries. It goes like this. Years ago, two tailors were strong rivals in a small town in Eastern Europe. Over the years, their rivalry became more extreme, developing into a hatred so intense that it came to the attention of the Lord Himself, who sent an angel to one of the rivals with an offer. The angel said, "Schmuel, the Lord has become aware of your rivalry with Yitzhak there, across the street, and has empowered me to settle this rivalry, once and for all, by offering you a wish." "One wish?" said Schmuel. "Anything I want?" "Yes," said the angel, "anything at all, but with one proviso: whatever you ask for, Yitzhak gets twice as much." Without hesitation, Schmuel pointed to his face and said, "Put out one of my eyes!"

In legends, wishes often lead to calamity. "Be careful what you wish for, you may receive it," so the saying goes. In stories where there are three wishes, the last wish inevitably causes major problems for the wisher. The Djinn and Leprechauns never allowed mere mortals to achieve their wishes at simple face value. The lesson was that mortals didn't have the insight or pure motives to appreciate the broad ramifications wrought by fulfillment of their expressed desires.

I am so interested in wishes that I conducted an international study of human desire in which I asked respondents to imagine the following scenario: You are walking down a path and come to a colorful stone. As you pick up the stone, you hear a voice saying, "Squeeze the stone with all your might, and your wishes will be granted. Use the stone as often as you like." You hold the stone in your trembling hand for a few seconds, close your eyes, and make your wishes. What are your wishes?

When I was a small child, I often found myself wishing for various physical powers, like those exhibited by such superheroes as the X-men or Superman. I sometimes wondered if I was the only one to do so, but haven't we all wished for a genie in the bottle? How many of us have made a wish while pulling on the wishbone of a chicken or while watching reruns of *I Dream of Jeannie*?

Since ancient days, people consulted priests, shamans, or other wise men for dream interpretation and wish-fulfillment. More recently, people consult psychic channelers, crystals, UFOs, and an amazing array of New Age paraphernalia. It seems humans have always wished for material possessions and spiritual powers. What is the significance of our specific wishes? How do our wishes changes with age? How do they vary with gender and culture?

I think of "wishing" as part of a whole cultural picture; people's wishes mirror their feelings and position in the rest of society. My experience reading and listening to thousands of people's wishes has made me realize that wishes are not casual but rather are rooted in the wisher's present life and concerns. In fact, it seems that wishes often replay people's lives in depth, dredging dreams that are almost subconscious until written down. You can read many of the wishes I collected at Pickover.com. As you'll see, a wish can give both literal information and also symbolic information revealing a person's inner world with all its conflicts.

My unpublished *Book of Wishes* was meant to keep our fingers on the pulse on the world, to eavesdrop into the usually hidden side of human desire as people express in their own words their inner fears and hopes. I've found the experience of receiving wishes from around the world to be enriching, warming, and enlightening, and I hope you will share in some of the pleasure when reading the wishes.

In my own life, I have found wishing for certain goals to be important. Having always desired to be a writer, I am now the successful author of more than thirty books and the associate editor of various scientific journals. My Internet Web site has received more than a million visits. Having "wished" to get an advanced degree, it took me three years to graduate first in my class in college, and I then quickly obtained a Ph.D. from Yale University. You, too, can achieve your own goals and dreams.

Imagine a future world where a kindly and wise Goal Giver assigns children fascinating goals that must be achieved in their lifetimes. When you are born, your parents are handed a list with one hundred goals. Some goals are difficult to achieve (pass a course on differential geometry and topology), while others are simpler (play "Silent Night" on the piano). As a stimulus to a nation's citizenry, if you achieve all 100 goals, there is a reward of one million dollars. What would such a world be like? What are some goals that a Goal Giver should assign? What would a human faced with this list really achieve?

Such an idea is not preposterous; in fact, there is a man today who forced himself to achieve over 100 goals set down on paper in the early years of his life. The man's name is John Goddard. When John was only a teenager, he took out a pencil and paper and made a long list of all the things he wanted to achieve in life. He set down 127 goals. Here is a list of just some of his goals:

1. Explore the Nile River
2. Play "Claire de Lune" on the piano
3. Read the entire *Encyclopedia Britannica*
4. Climb Mt. Everest
5. Study primitive tribes in the Sudan
6. Write a book
7. Read the entire Bible
8. Dive in a submarine
9. Run a five-minute mile
10. Circumnavigate the globe
11. Explore the Great Barrier Reef of Australia
12. Climb to the very top of Cheops's Pyramid

Impractical? Not at all. John Goddard has accomplished more than a hundred of his original 127 goals. He's become one of the most famous explorers in the world. Goddard is the first man in human history to explore the entire length of both the Nile and Congo rivers.

When I read about John Goddard in early high school, I made my own list. Admittedly, many of these are fanciful, the product of a teenage mind:

1. Play Bach's "Toccata and Fugue in D Minor" on the piano (achieved 1990)
2. Learn Ch'ang-Shih Tai Chi and Shaolin Kung Fu (achieved 1996)
3. Obtain a Ph.D. from Yale University (achieved 1982)
4. Sell a novel (achieved 1995)
5. Raise golden Amazon sevrum fish (achieved 1990)
6. Play bass guitar in a rock band (achieved 1975)
7. Eat spicy tekka maki (achieved 1994)
8. Own a Mitsubishi 3000 sports car with a stick shift (achieved 1994)
9. Fire an Uzi submachine gun and Magnum .45 (achieved 1993)
10. Publish a technical paper with a triple integral symbol (achieved 1986)
11. File a United States patent (achieved 1986)
12. Have a book published in Japanese (achieved in 1991)
13. Have a professional massage (achieved 1993)

14. Have a book turned into a movie (not yet achieved)
15. Read Will and Ariel Durant's entire *Story of Civilization* (not yet achieved)
16. Visit St. Peter's Basilica (achieved 2004)
17. Eat fugu (not yet achieved)
18. Find *Adventures of a Grain of Dust*, a book lost since childhood (achieved)
19. Learn to play the Japanese game Go well (not yet achieved)
20. Fully understand the concept, "All that is not given, is lost" (not yet achieved)

In contrast to the broader selection of wishes in *The Book of Wishes*, my list and Goddard's are all achievable. If you make your own list, remember that all the wishes on your list are important. Like Goddard and me, you will become a happier person by accomplishing your goals. Our world needs happy, satisfied, self-assured people to provide humanity with ideas and inspiration.

Before soliciting thousands of wishes from around the world, I had thought that most people would wish for the impossible, such as immortality or the power to move objects with their minds, but to my surprise the overwhelming majority of wishes fell into achievable and possible goals. Examples of "possible" wishes include finding a mate, gaining money or fame, or having safe abortions available to the world. I was surprised how few respondents mentioned religion, asked to meet God, or to know if God exists. I found that the wishes of female respondents dealt with family matters twice as often as did men's wishes. Men were more interested in wishes dealing with the intellect and knowledge than were women. People younger than 40 were more concerned with jobs, spiritual matters, and desire for power and knowledge, than people older than 40. In my wish survey, women were more interested in pets and animals than men.

Many of us are too shy to express our wishes. So be it—but let's not grow hardened to our own secret wishes. Knowing and expressing dreams and wishes is first step to realizing them. Your wishes are your muses, your sources of inspiration. In his famous "I Have a Dream" speech in 1963, Martin Luther King Jr. moved an entire nation to aspire to his vision of freedom and equality for all.

"Work only a half a day; it makes no difference which half—
it can be either the first 12 hours or the last 12 hours."
                                    —Rule 1 in Holiday Inn founder's
                            "Kemmons Wilson's Twenty Tips for Success"

"Mental attitude plays a far more important role in a person's
success or failure than mental capacity."
                                    —Rule 3 in Holiday Inn founder's
                            "Kemmons Wilson's Twenty Tips for Success"

"The secret of happiness is not doing what one likes, but in
liking what one does."
                                    —Rule 7 in Holiday Inn founder's
                            "Kemmons Wilson's Twenty Tips for Success"

"I got up from the table and realized, on the way to the couch,
that everything I knew was based on a false premise. I fell
down through the couch into another world....I noticed that
time did not extend smoothly—that it was punctuated by
moments—and I fell down into a crack between two moments
and was gone...[Later] I remember sitting on a bench, waiting
for a class to begin, thinking 'That was the most incredible
thing I've ever done.'"
—Nobel Laureate Kary Mullis, *Dancing Naked in the Mind
                    Field* (description of an LSD altered state)

"I can envision devoting a single day in the near future on which, say, five million people worldwide took a good healthful dose of MDMA (or hashish, psilocybin...) and opened up their hearts and minds to each other and to the universe. Such a rite of pure Dionysian grace, involving communal song, dance, and invocations of prayer, would strum the invisible wires of the emergent global consciousness network, striking a harmonious chord from Chicago to Bangkok, Sydney to Sao Paolo, London to Delhi, Durban to Tehran."

—Charles Hayes,
"Is Taking a Psychedelic an Act of Sedition?" *Tikkun*

The Business of Book Publishing: Unplugged, Up Close, and Personal

In which we encounter prolific writers, literary agents, publishers, editors, bibliomaniacs, "The Whore of Mensa," Hillary Clinton, William Shatner, coffee-addict Honoré de Balzac, Dr. Seuss, cartoon guides to Proust, Isaac Asimov, publisher response times, famous rejected books, authors' advance money, lecture tours, book signings, book dedications, literary agent John Brockman, collaborations, Piers Anthony, Shirley Jackson, ghostwriters, book titles, cover art, bestsellers, writing advice, Lester Dent, Erle Stanley Gardner, Dean Koontz, Gertrude Stein, Truman Capote, Perry Mason, Willa Cather, Sigmund Freud, Ring Lardner, John Creasey, Spider Legs, Pycnogonids, Jerzy Kosinski, Ellery Queen, John F. Kennedy, Agatha Christie, John Cheever, Alfred Knopf, and amphioxus.

"A book is like a garden carried in the pocket."[1]

True Journey Is Return

Readers frequently ask me about the methods I have used to become a prolific book author. Readers also wonder how they can get published. So, in this chapter, I would like to talk about personal experiences I've had as a writer and with the business of publishing. In the previous chapter, I mentioned that Marcel Proust was forced to pay for the publication of his masterpiece, *In Search of Lost Time*. Luckily, I've not had to do that with any of my books. In contrast, no publisher would initially touch *In Search of Lost Time*, even though today it is hailed as one of the best novels ever written. In 1919, one of its earlier volumes, *Within a Budding Grove,* won France's most prestigious literary award, the Goncourt Prize.

We can see why I can easily find a publisher for a majority of my books and Marcel Proust could not. Most readers today never finish Marcel Proust's 3,000-page *In Search of Lost Time*. It's essentially several different novels rolled into one. I personally have never known a single person who has finished reading it. Some friends tell me that they have read as far as Swann's marriage. Others say that they gave up at the 200-page party scene. Many friends enjoy reading the long scenes describing Proust's difficulty falling asleep to help them fall asleep! The best most mortals can do is read the dozens of "how-to-read" Proust books. I've read these myself, and I also highly recommend the cartoon guides and graphic novels for *In Search of Lost Time* created by Stephane Heuet.[2] Most people can probably get through the novel condensed to a comic book.

Despite Proust's serpentine sentences, many people do revere his work and use his name as a symbol of genius. Woody Allen's short story "The Whore of Mensa" describes a man who craves the company of a brilliant sexy woman.[3] The man says that he can meet all the "bimbos" in the world, but what he really wants is a young woman who, for a price, will come to his bedroom to discuss Proust with him!

Proust apparently had a darker side to his life, with which most people are unfamiliar. For example, according to French writer Maurice Sachs (1906–1945), Proust achieved sexual release by watching pins stuck into live rats.[4] He also seemed to get sexual enjoyment by defiling objects of veneration, for example, by spitting on or insulting photos of his mother, whom he loved.[5] These weird anecdotes may not be reliable, but certainly Proust was not your normal guy. He admitted he was strange and said, in *The Guermantes Way*, "All the greatest things we know have come to us from neurotics. It is they and they only who have founded religions and created great works of art. Never will the world be conscious of how much it owes to them, nor above all of what they have suffered in order to bestow their gifts on it." So, despite Proust's eccentricities, I admire his dedication as a writer, his talent, and his limitless intellect.

The essence of *In Search of Lost Time* can be described in just a few words: A man leaves home, searching for happiness. He does not find happiness by traveling, but instead learns that happiness comes from within and in the home he left behind. True journey is return. Another Proustian message: Slow down. Take life at a more leisurely pace. Take your time.

Prolific Writers

I have never really taken Proust's "slow down" message to heart. I'm one of the most prolific authors in the world today, if you consider not only the number of my books published since 1990, but also the range of topics, which include mathematics, religion, history, science fiction, psychology, astronomy, physics, medicine, computers, and even divination. When people ask me how did I do it, I reply: "Some people play golf on the weekends. Instead, I prefer to write." Of course, my prolific writing pales in comparison to American novelist, lawyer, and workaholic Erle Stanley Gardner (1889–1970), who once worked on seven novels simultaneously and dictated 66,000 words a week. Gardner would never start to dictate until he had worked out the entire plot of his novel. He actually hired six secretaries to handle his dictation, which he found more efficient than typing. His best-known works focus on the lawyer-detective Perry Mason.

The actual act of writing is something that writers seem to enjoy. The pleasure we feel was expressed eloquently by U.S. historian, journalist, and novelist Henry Adams:

> The fascination of the silent midnight, the veiled lamp, the smoldering fire, the white paper asking to be covered with elusive words; the thoughts grouping themselves into architectural forms, and slowly rising into dreamy structures, constantly changing, shifting, beautifying their outlines   this is the subtlest of solitary temptations, and the loftiest of the intoxications of genius.[6]

In 1885, Adams's wife committed suicide, and he suddenly became a compulsive traveler. He had already been to Japan and Cuba, but in 1890 he embarked on a grand journey across the globe to Hawaii, Samoa, Tahiti, Fiji, Australia, Java, Ceylon, Aden, Paris, and London. He may have seen more of five continents and of the United States than any other American of his era.

I don't know how writers like Isaac Asimov were so prolific before the age of the computer. I would have a very difficult time writing books, and doing all the necessary text rearrangements and editing, without a word processor. Asmiov had a U-shaped desk and three typewriters in his office. If he ever became bored or stuck on one project, he simply swiveled his chair and worked on another project.

According to the *New York Public Library Desk Reference* (4th ed.), Isaac Asimov wrote over 400 books and is the only author with a book included in every major Dewey-decimal category. I sit in awe of Asimov, but a few people have exceeded his book output. Lauran Paine (b. 1916) has published over 900 books under more than 90 pen names. Paine spent his youth working as a cowboy, and today at least 500 of his books are Westerns.

Authors of the past used a variety of techniques for starting a book and had a host of interesting writing customs. For example, Willa Cather, Nebraska's most noted novelist, had to read a line from the Bible before she started her writing for the day. Cather won the Pulitzer Prize for her novel *One of Ours*, about a Nebraska farm boy who went off to World War I. Lewis Carroll and Virginia Woolf often wrote while standing. Other authors were very particular about the colors of their inks and papers. For example, Rudyard Kipling hired an "ink boy" to grind extremely black India ink.

American author Truman Capote had this advice for young writers: "Socialize. Don't just go up to a pine cabin all alone and brood. You will reach that stage soon enough anyway." Honoré de Balzac (1799–1850), one of the greatest French novelists who ever lived, drank over 50 cups of black coffee a day to help him write. He published over 90 novels and wrote up to 15 hours a day. Regarding his coffee addiction, he wrote:

> For a week or two at most, you can obtain the right amount of stimulation with one, then two cups of coffee brewed from beans that have been crushed with gradually increasing force and infused with hot water. For another week, by decreasing the amount of water used, by pulverizing the coffee even more finely, and by infusing the grounds with cold water, you can continue to obtain the same cerebral power.[7]

In the end, Balzac ate dry coffee grounds to achieve the desired stimulant effect. Alas, caffeine poisoning was a major contributor to his early death.

Marcel Proust was also a big coffee fan, and he required the brew to be thick and as strong as possible. His stimulant abuse, which included caffeine and adrenaline, required him to take opium at bedtime to calm down.[8]

The Difficulty of Selling Fiction

I have published both fiction and nonfiction books, but finding a publisher for fiction is much more difficult than for nonfiction. As just one example, I recently employed a literary agent to sell my novel dealing with technology and the afterlife. She was a former publishing ex-

ecutive and also a writing teacher. Alas, she couldn't get a single potential publisher to respond to her queries about the novel. The novel was
probably quite marketable, but she wasn't able to get a publisher to return her phone calls or e-mails. Perhaps this says a lot about the agent's
lack of appropriate contacts, but, more generally, if the ability to find a
publisher for nonfiction can be compared to walking across the street,
finding a publisher for fiction is like walking from New York to California, backward. According to Marc McCutcheon, author of *Damn! Why
Didn't I Write That?*, of the 50,000-plus new books published each year,
only about 3,500 are fiction. A mere 120 fiction releases each year are
first novels of an author.

I'm not the only author to realize the immense challenge in publishing fiction. For example, John Scalzi (Scalzi.com) has written extensively on the difficulty of publishing science fiction. First let me tell you
a little bit about John. He is a published nonfiction book author. He has
written for the *Chicago Sun-Times*, the *Washington Post*, the *San Diego
Tribune* and the *San Francisco Examiner*. He's even been on *Oprah*.
John has been an editor, most notably for a humor area on America Online.
He even has an agent—for his nonfiction.

Scalzi believes that in order to get science fiction published, you need
to have accomplished only two things: (1) You must have published a
science-fiction novel in the past, and (2) you must be writing *military
science fiction*.[9] He's conducted extensive surveys of bookstore shelves
devoted to science fiction. By his estimate, eight out of the ten books on
the shelves are from well-established science-fiction writers, many of
them dead, such as Asimov, Clarke, Heinlein, and Niven. The remaining
two books of the ten are from relatively new authors, and most of these
are "hard" science fiction dealing with space battles, military machines,
starships, and the like. Scalzi concludes that this leaves only "four science fiction books out of a hundred that feature new authors *not* writing
about space navies and powered war suits."[10]

I think Scalzi is largely correct. For example, I did publish a novel
Spider Legs with the big science-fiction publisher TOR, but TOR published it because I had collaborated with science-fiction and fantasy superstar Piers Anthony, who has published over a hundred novels. When
I attempted to publish my four books in my Neoreality series, none of
the major publishers was interested, and so I had to publish the series
with a small publisher. Happily, the books were immediately and favor-
ably reviewed in *Analog* magazine, one of the premiere science-fiction

magazines in the world. You'll learn more about the Neoreality series in Chapter 8.

Pen Names, Rejections, Dr. Seuss, Silence of the Lambs

I have never written a novel under a pen name, but was tempted to when Oxford University Press, one of my science book publishers, thought I was writing too many science books and suggested I slow down. English author John Creasey wrote 564 books under 13 pen names; however, he received 743 rejection slips before a publisher accepted one of his mystery novels! James Joyce's *Dubliners* was rejected 22 times. Anais Nin, Agatha Christie, Sinclair Lewis, and Ernest Hemingway were rejected many times—even Stephen King has had many manuscripts rejected. Classic best sellers like *War and Peace*, *The Good Earth*, *To Kill a Mockingbird*, *The Rubaiyat of Omar Khayam*, and *Watership Down* were all rejected by numerous publishers.

The following information may be slightly exaggerated, but everyone I meet tells me that Dr. Seuss's *To Think That I Saw It On Mulberry Street* was rejected 24 times, *Silence of the Lambs* 28 times, *Catch-22* 21 times, and *Zen and the Art of Motorcycle Maintenance* more than 100 times. Twenty publishers felt that Richard Bach's *Jonathan Livingston Seagull* was for the birds. It went on to sell millions of copies around the world.

As for *Sex, Drugs, Einstein, and Elves*, I submitted a book proposal to numerous publishers, and in each case I targeted the proposal to a particular nonfiction editor whose interests were likely to intersect some of the topics in this book. The following publishers rejected the book with a form letter: Atria Books, Warner Books, Newmarket Press, Thames & Hudson, Pelican Publishing, Red Wheel-Weiser-Conari Press, Coffee House Press, Verso, and Rutgers University Press. Most form-letter rejections indicated the publisher would not look at unsolicited proposals or proposals submitted without an agent—or that the book was not appropriate for their publishing program.

The following publishers rejected with a personal letter, or at least the illusion of a personal letter: Adams Media, Princeton University Press, Kensington Publishing Group, Prometheus Books, McGraw-Hill, Chronicle Books, Andrews McMeel Publishing, Simon & Schuster, Johns Hopkins University Press, Berkley Books, HarperCollins, University of Chicago Press, Cornell University Press, University of California Press, Vintage, MIT Press, BenBella Books, NYU Press, Knopf, Beacon Press,

Soho Press, Routledge, Workman Publishing, Oxford University Press, Harvard University Press, Harcourt, and Tarcher/Penguin.

Personal rejections for *Sex, Drugs, Einstein, and Elves* tended to contain uplifting fragments, including such niceties as: "This is truly one of the most unique ideas I've seen in a long time..." (Adams Media), "I enjoyed reading your new proposal and am almost ready to move to Shrub Oak" (Princeton University Press), "We appreciate your surrealistic take on culture, and your no holds barred approach to a memoir" (Chronicle Books), "I'm sure there will be others who are fascinated by the associations and connections that you make" (Simon & Schuster), "I like what I read" (Johns Hopkins University Press), "The material is certainly interesting—and in some strange way, I quite enjoyed reading it (Berkley Books), "I found it to be quite a ride..." (Penguin Books), "Lots of fun...Ideas fizzing all over the place..." (Cornell University Press), "I was honored to receive a copy of your proposal for *Sex, Drugs, Einstein, and Elves*...I have followed your publishing activities over the past 15 years..." (MIT Press), "It looks intriguing..." (Knopf), "I love your work" (Beacon Press), "I bet this book will get snapped up by someone" (BenBella Books), "I really enjoyed reading it, by the way" (Oxford University Press), "...It is a very interesting project..." (Yale University Press), "I share the enthusiasm of the other readers for your rather quirky musings....Very enjoyable reading" (Stanford University Press), and "The book is a close encounter with your amazing mind...and a *tour de force*" (Harvard University Press).

I submitted this book to about 80 publishers. All rejected it except for the current publisher. The first rejection letter came on August 2, 2004, and the acceptance letter came on October 4, 2004. The current publisher sent the contract as a Word document through e-mail. I returned the contract with some modifications, and then we signed. The contract states that I will receive 15% of the gross sales, that is, money received by the publisher for sales of this book. I will receive 50% of the amount received for the disposition of secondary rights such as translation or motion picture rights.

I advise authors to list excerpts from publisher rejections in their books because this gives the writer something to look forward to with each rejection as he or she searches for an interesting textual nugget for inclusion in the book. My favorite letter came from Columbia University Press, who eventually rejected the book:

Dear Dr. Pickover,

We have never met, but reading your *Sex, Drugs, Einstein, and Elves* outline and sample, and investigating you in more detail, makes me feel that I did. Your descriptions of Shrub Oak, which I must admit I had never really seen as a nirvana, now makes me wonder whether choosing to live in Cortlandt Manor was wise when Shrub Oak was just down the road.

Anyway—I did enjoy the bit you sent me about *Sex, Drugs, Einstein and Elves*, and I would like to see more of the manuscript if you feel you can show it to me. I admire your chutzpah in using the Princeton University Press quote in your submission. I wish I could be confident that Columbia is radically different and would embrace your project with alacrity—quite honestly I just don't know, so maybe seeing more would help?

Perhaps all this is moot as maybe a fat advance has already arrived at Shrub Oak from some fine publisher for this book? Please do let me know. I *am* interested in the book, but really feel I must read more to decide. Can you help?

Today's editors or editorial assistants have rejected such classics as *War and Peace, Moby Dick*, and *Gone with the Wind*, when these novels were submitted untitled as a way to test the submission process. Author Chuck Ross recently retyped *Steps* by Jerzy Kosinski. This book was originally published by Random House and was both a best seller and National Book Award winner. Next, Ross sent the book to roughly 30 literary agents and publishers—including Random House, the original publisher who sold some 500,000 copies of the book. All 30 publishers and agents rejected the work, including Random House.

Some fiction publishers receive so many submissions in just one day that they would need huge staffs just to read the first few pages of each submission. Even having an agent can be fruitless, as I have found for my own fiction. For example, publishers will read few manuscripts submitted by agents who are not very well connected or well known. Many publishers and authors have said, "Your odds of getting hit by lightning are higher than getting a novel published by a name publisher."

Some Web sites actually track the amount of time it takes a publisher to respond to a science-fiction book or magazine article submission. Dr. Andrew Burt, for example, publishes a huge list of publishers and their average response time. He writes, "When writers submit a manuscript to an editor, they often feel like they've launched it toward a black hole."[11] Here are some of Burt's average response times for a sampling of publishers: Avalon Books (587 days), Avon Eos (294 days), Baen (329 days), and TOR (115 days).

Some writers test to see if publishers even glanced at the manuscripts before rejecting them. Editor Walter Hines Page (1855–1918) received the following letter from a contributor, "Sir: You sent back last week a story of mine. I know that you did not read the story, for as a test I had pasted together pages 18, 19, and 29, and the story came back with these pages still pasted; and so I know you are a fraud and turned down stories without reading them."

Page replied, "Madame: At breakfast when I open an egg, I don't have to eat the whole egg to discover it is bad."

You'll find that all the "how to" books for getting your book published insist that you send a self-addressed stamped envelope (SASE) with each submission. This piece of advice always amused me. Why bother? If an editor likes your submission, do you think he or she will care about an SASE? And if the editor wants to reject it, I don't really need to hear about it. In my cover letters, I tell the publishers they may discard the submission if uninterested. I always send submissions to several publishers simultaneously, unless a publisher has contacted me first about a specific book idea.

American author Ring Lardner (1885–1933) gave this advice to a younger writer: "Don't make the mistake of enclosing a self-addressed envelope big enough for the manuscript to come back in. This is too much of a temptation for the editor."

The Author "Advance" Money

Marcel Proust dreamed of being a novelist as a little boy, and he had numerous false starts through his life before he wrote *In Search of Lost Time*. It took me a year or two to interest a publisher in my first book, *Computers, Pattern, Chaos and Beauty*. The first book is often the hardest to sell. I recall that some publishers said they wanted it, but then changed their minds as we got close to signing a contract. One publisher had me actually sign a contract, and then backed out. In some ways these initial turndowns were good luck in disguise, because I finally signed a contract with St. Martin's Press, who offered me an advance of $5,000, which was actually higher than the previous publishers' proposed advances. The term "author advance" refers to the amount of money a publisher pays an author before the book is available in stores. The author typically gets no more money from the publisher until the book earns royalties exceeding the advance. I've had several advances greater than $30,000 but no six-figure advances. *Computers, Pattern, Chaos and Beauty* went on to sell more than 25,000 copies, very nice for an author's first book.

The advance has two major purposes. One is obviously to ensure that the author gets a certain amount of money. Often, an author never gets a dime more than his advance. A second purpose is to ensure that publishers devote some effort to promoting the book so they can at least get back their advance money. A significant advance would seem to guarantee a certain level of commitment to the book. Dean Koontz has noted that publishers push hard with advertising and promotion only on those books on which they were at risk of losing money.

Interestingly, in 1899, Sigmund Freud was paid only $209 for his famous *The Interpretation of Dreams*. After its publication, just 600 copies were sold in eight years, yet it later became a landmark book. Edgar Allan Poe received $10 for his famous poem "The Raven."

Because a person can't live on advances like these, many famous authors also had other less-glamorous jobs before their books became major sellers. Henry Miller dug graves for a living. William Burroughs, author of the cult classic *Naked Lunch*, was an exterminator in Chicago and a bartender in New York.

Advances also often come in two parts, half upon signing the book contract, the other half upon acceptance of the final manuscript or publication of the book. This two-stage advance has led to interesting speculation about the prolific American short-story writer O. Henry. He was a master of surprise endings and often wrote about the life of ordinary people in New York City. The *New York World* paid him $100 for each of his stories, but because O. Henry did not always deliver his story by the deadline, his editor made sure only to pay half of the advance for the first half of the story and refused to pay the rest until the entire story was received. Some critics have said that this led to O. Henry stories with the second half nearly unrelated to the first! Apparently, O. Henry sometimes rushed out anything he could for the first half of the story just to get the $50 and then ignored much of what he had written when completing the tale.

Literary Agents

I mentioned that my first book, *Computers, Pattern, Chaos and Beauty* was one of my better-selling books. Note, however, that my *Mind-Bending Puzzles* calendars, published each year, outsell my books. I stayed with St. Martin's Press for the next three books, reaching one of my highest cash advances for *Mazes for the Mind: Computers and the Unexpected*. The $42,000 advance was achieved with the help of literary agent John Brockman, the world-renowned agent for scientist writers.

The concept of literary agents is an interesting one. Many successful writers have managed just fine without them. Isaac Asimov's last agent left the business in 1953, and Asimov sold hundreds of book over the next forty years without an agent. I believe Asimov once said that he didn't like working with literary agents because they placed a layer between him and the publisher. He likened the process to kissing a girl through cellophane. Remove the cellophane, and it feels much better.

I've had a few agents, and with the exception of the superstar John Brockman, I have found agents to be essentially useless. Most agents have simply contacted the publisher for my previous book, with whom I was already intimate, and secured deals I could have gotten on my own. Thus, I decided to sell most of my books without an agent. Some agents I used were so poorly connected to publishers that the agents actually asked *me* to provide them with a list of potential publishing contacts, editor's names, e-mail addresses, and then the agents relied exclusively on my list. Other agents wanted me to sign contracts that would commit me to use them for future books. However, this seemed unreasonable to me because I wanted to first determine how well we worked together on a current book and to also see if they were capable of getting me deals that were better than I could get for myself. Through my career, it seemed to me that the best agents were those who didn't demand that I use them for future books.

Even though some agents will have more credibility with publishers than you do, if you are a beginner, for the most part, agents will not pitch your book as aggressively as you would. Many dribble the book to publishers one or two at a time. Why not twenty? What's there to lose by hitting many publishers at once? That's what I do, and I've always seen an advantage to this approach.

Finding an agent can be as difficult as finding a publisher. And some agents can be rather disorganized. For example, after I had found an agent for one of my novels, I e-mailed two other agents who had not yet decided if they wanted to represent me, but who had the book in their hands. I asked them if they could make a decision because I was about to contract with the interested agent. The two agents advised me to sign with the agent whom I had found. A month later, the same agents who had rejected me sent me a contract for representation in the mail. That's fairly disorganized—turning me down and a month later offering me a contract!

As I mentioned, the most effective agent I ever had was John Brockman, a New York literary agent who handles authors writing about cutting-edge science and how it is changing the world. His Web site, www.edge.org, has a cult following and is both a magazine and online community.

To show you how selective Brockman can be for the books he decides to represent, I had a book deal ready to go with Wiley for my book *A Passion for Mathematics*. I asked John if he wanted to handle the contract work and get his cut of the royalties and handle foreign-rights deals, and he declined because (I think) he didn't feel that the book would have wide enough appeal. Now, that's pretty selective, given that I already had a publisher interested and offering a $15,000 advance. He also was not interested in representing *Sex, Drugs, Einstein, and Elves.*

But a good agent, like Brockman, can do much more for an author than simply get the author a good advance. He can also get good press for the author through his Web site and many articles and books that feature his favorite authors. As just one example, Brockman got my name into *The New York Times* after they mentioned a piece I had written for him. He also had me contribute to some of his books of essay collections.

Collaborative Novels

Many readers have asked me how two authors can successfully work together on a novel. One of my favorite collaborations was with Piers Anthony, the well-known fantasy novelist. Although my popular science books gave me a nice sense of accomplishment, in the 1990s my real dream was to publish a novel based on my interest in unusual aquatic creatures, particularly pycnogonids, or "sea spiders." The book *Spider Legs*, a Pickover/Anthony collaboration, was based on my explorations, on land and in the sea, into the rare and dangerous creature known as *Colossendeis*. Yes, the deep-sea *Colossendeis* in the novel is real! Pycnogonids are real. If you ever read *Spider Legs*, note that the various biological descriptions in the novel, such as the packing of the pycnogonid's digestive system into its legs, are based on scientific facts. However, the life cycles of the large, deep-sea forms, especially members of the genus *Colossendeis*, are still largely unknown to scientists.

How does Piers fit into all this? After completing a rough draft of the novel, I began to search for a collaborator to help polish the novel and add material to deepen the character's motivations. My first thought was Piers Anthony, science fiction and fantasy's most creative talent—and one of the most prolific. He's had several *New York Times* best sellers. I had been reading his books for many years, but the idea to collaborate with Piers started when a colleague lent me a copy of Piers's fantasy novel *Virtual Mode*, which had just been published. To set the stage, I mailed Piers my book *Computers and the Imagination*, and I thought this would prepare him to receive further material from me. I waited a week or two. Then I followed up by sending him a draft of *Spider Legs*, which at that time was probably only about 40,000 words.

After some hesitation on Piers's part, it seemed like I soon hooked him on the idea of a collaboration. Collaborating turned out to be quite easy, and oddly enough, choosing a title was one of our more difficult jobs. I had originally called the book *Phantom*, a title we abandoned because it had been used too many times. Before we finally arrived at *Spider Legs*, we considered other titles like *PycnoPhantom, Legs, Killer Legs, Sea Legs, Pyncophobia, Fractal Phantoms, Spider Eating, Spider Hunter*, and even *20,000 Legs Under the Sea*. The collaboration was relatively painless. He doubled the size of the book, focusing a great deal of attention on character development and enhancing the dialogue. We were nearly done in a month.

Collaborations are quite common for novels, although that collaboration is sometimes shielded from the public. For example, cousins Frederic Dannay and Manfred B. Lee worked together on hundreds of mystery novels and stories as "Ellery Queen." For some time, "Ellery Queen" was marketed as a secret identity. Dannay would appear as Ellery Queen in public masked, as though he were protecting his identity. The buying public loved it.

Later, the cousins created a new author, "Barnaby Ross," whose existence had been foreshadowed by two comments in the Queen novels. Barnaby Ross published four novels, at which point the publisher revealed that Barnaby Ross was actually Ellery Queen (who didn't really exist)!

One of the largest author collaborations was the mystery novel *The Floating Admiral*, written by Agatha Christie and twelve other writers. One author would write a chapter, hand it to the next author, who would add the next chapter, and so forth, until the book was finished. The novel starts with the floating admiral of the title found sailing along in an abandoned boat with a dagger in him. Dorothy L. Sayers explained the writing ground rules in the introduction to the book: "Each writer must construct his installment with a definite solution in view—that is, he must not introduce new complications merely 'to make it more difficult'...Each writer was bound to deal faithfully with all the difficulties left for his consideration by his predecessors."[12]

Book Ghostwriters

A ghostwriter is an author who writes a book for another person whose name goes on the book. Thus, the author on the cover is not the actual writer. Sometimes publishers employ ghostwriters to write a book for celebrities like presidents and movie stars who may not have the time or skill to actually write a full-length book. For example, *Profiles in Courage* was a famous ghostwritten book, which has John Kennedy's name on it. Ghostwriters usually receive no public credit for creating the work.

Before it became known that *Profiles in Courage* was largely done by a ghostwriter, it was awarded the Pulitzer Prize for biography. Perhaps President Kennedy was the first person to win a Pulitzer Prize for a book written by someone else. Of course, we shouldn't be too harsh on President Kennedy, because even critics agree that he assisted with many choices, messages, and the overall tone of the book.

My favorite ghostwriter anecdote involves Christy Walsh, Babe Ruth's business manager. The ghostwriter for Walsh was sportswriter Tom Meany. At one point, Meany was so busy that he had to hire a ghostwriter to do his ghostwriting. A ghost of a ghost!

According to Mahesh Grossman, author of *Write a Book Without Lifting a Finger*, Pamela Anderson, Hillary Clinton, and Dr. Phil all used ghostwriters. Grossman says that William Shatner has a dedicated group of four doing it and that almost half of all published authors use ghostwriters.[13] Grossman asserts that there is nothing to be ashamed of when an author uses a ghostwriter, because the author usually provides expert knowledge and thoughts, which the ghostwriter translates into professional prose.

Book Titles and Cover Art

I generated almost all of the titles of my books on my own, without help from the publisher, and most people tell me they like the titles quite a bit. Sometimes, when I am undecided about a title, I send a few possibilities out to friends for their assessment, and I use the title that receives the most votes. Some favorite Pickover book titles are *Keys to Infinity*, *The Lobotomy Club*, *The Mathematics of Oz*, *Surfing Through Hyperspace*, *Sushi Never Sleeps*, *Liquid Earth*, *The Alien IQ Test*, *Computers and the Imagination*, *Chaos in Wonderland*, *Calculus and Pizza*, and *The Loom of God*.

Many of my publishers have involved me when making decisions about the cover art of my books, and I actually designed many of my own covers using computer graphics. The covers on which I had the most input often turned out the best. Occasionally, I am amazed and shocked at how poorly some covers turn out when designed by the publisher, and it's quite insane that a publisher can't see a poor cover for what it is. For example, the cover of my book *The Science of Aliens* could have been amazing, featuring some eye-catching alien, but the publisher chose a washed out, unattractive cover, with almost no contrast between the lettering and the background puke color. The book title is completely unreadable on the book's spine because the letters are white on top of a light green background. Clearly, no human being could ever read the book's title on the spine. So how is it that an experienced publisher could actually design a cover so badly, especially after investing so many thousands of dollars in the book? This is a question I have never been able to answer. I was also amazed that the cover for *Spider Legs* features a giant lobster that appears nowhere in the book! However, aside from one or two bad covers, I've been delighted with my book covers.

I know that authors, even the biggest book authors, often agonize with their publishers over imaginative titles that will maximize book sales. Dean Koontz, for example, often debated back and forth with his editors about book titles. I recall that his publishers often wanted one-word titles, so, for example, they changed Koontz's *Lighting Road* to *Lightning*. Koontz is certainly the master of one-word titles with his bestsellers like: *Midnight*, *Shattered*, *Icebound*, *Phantoms*, *Whispers*, *Strangers*, *Hideaway*, *Ticktock*, *Intensity*, *Shadowfires*, and *Watchers*.

Editor Malcolm Cowley (1898–1989) once told his staff, "Let's have a big novel written to order, and let's give it a tremendous title—you know, something that will make people think of *The Good Earth* or *All*

Quiet on the Western Front. Any bright ideas?" One of the company's advertising managers replied, "What about *All Noisy on the Eastern Behind?*"

Isaac Asimov was always careful about his titles, preferring short titles to long titles. When possible, he preferred to have one-word titles for his books and stories, such as "Nightfall" or *Foundation.* As Asimov describes in his autobiography *I. Asimov: A Memoir*, his editors were always tinkering with his title ideas. He wanted to title his book of short biographies of scientists, explorers, and inventors *A Biographical History of Science.* His publisher insisted on adding "and technology," although Asimov thought it was unnecessary. His editors also believed that the word "history" would decrease sales and wanted the word "Encyclopedia" in the title, although Asimov objected to this misrepresentation. Finally, the editor added "Asimov" to the title, thus creating *Asimov's Biographical Encyclopedia of Science and Technology.* Asimov concludes, "I must admit I swallowed the clumsiness of the title because of the first word."

I recently became so enamored with strange titles of published books that I ran a contest and asked colleagues: "What is the strangest, coolest book title you can find at Amazon.com?" Here are some of the suggestions: Andy Griffith's *Day My Butt Went Psycho*, Ariel Dorfman's *How to Read Donald Duck: Imperialist Ideology in the Disney Comic*, Hiroyuki Nishigaki's *How to Good-Bye Depression: If You Constrict Anus 100 Times Everyday. Malarkey? or Effective Way?*, P. P. Hartnett's *Mmm Yeah*, Chuck Klosterman's *Sex, Drugs, and Cocoa Puffs*, Geoff Dyer's *Yoga for People Who Can't Be Bothered to Do It*, and Steven Boyett's *Treks Not Taken: What If Stephen King, Anne Rice, Kurt Vonnegut and Other Literary Greats Had Written Episodes of Star Trek: The Next Generation*?

Tips for Writers: How to Write a Best Seller

Proust composed his books in a haphazard fashion. He did not start at the beginning and finish at the end. He did not write linearly. Instead, ideas came to him in flashes as he went about his daily routine. Most of my own books are composed in the same way. As ideas come to me during the day or in the realm between sleep and wakefulness, I jot them down and continue to fill in details throughout the book. For me, writing is exactly like painting, adding a spot of color here, a detail there, a twig on this tree, a bit of foam on that ocean wave…No painter starts at the top of the painting and finishes at the bottom.

My approach to filling in detail, like a painter dabbing paint, is fine in the age of word processors, but it was amazing that Proust used the same approach so well. He would dictate to his stenographers who would type an initial manuscript. Then he would crowd the margins with additional details and establish links between scenes and characters. He would paste in new pages and have the new work typed again and again. Edmund White notes in his biography of Proust, "If any writer would have benefited from a word processor, it would have been Proust, whose entire method consisted of adding details here and there and of working on all parts of his book at once."[14]

When I start writing my novels, I never know how the stories will end. I let the characters and initial crazy situation drive the plot, forcing both me and the book's characters to attempt to solve the challenges that come along.

After writing a few novels with good reviews, I feel I have some good advice to offer budding novelists. Let's first go over some of the mechanics. If you follow these suggestions, you will have a much better chance of selling your novel.

1. *Show Not Tell*

It is better to show through a character's actions than "tell" by having the narrator describe.

Example 1: "Garth was nervous" is "telling." It is better to "show" with: "Garth's hands trembled."

Example 2: "Garth did not want to go down the hall with the Sergeant" is telling. It is better to show with: "What?" Garth said, slamming his fist onto the table. "There's no way in hell I'm going with you!"

2. *Body Movement*

Occasional reference to body movement and setting is important so that characters are not disembodied talking heads. Body movement can sometimes be used before a person talks to establish who is talking.

"When are you leaving for France?" John asked.

This could be recast cast as:

John took a slow breath. "When are you leaving for France?"

3. *Short Better Than Long for Dialog*

In real life, people often talk in short sentences and phrases, so don't use long, drawn-out sentences with big words. Another dialog tip: use contractions often. For example, a character may be more apt to say "I'll" than "I will."

4. *Break the Dialog*

Always insert a "he said" or "she said" as early as possible into a line of dialog (if a "he said" is even needed at all).

Never do: "Yes, I will kill him, but not until you buy the large peaches for dinner and lace them with arsenic," he said. *Instead do*: "Yes," he said, "I will kill him, but not until you buy the large peaches for dinner and lace them with arsenic."

5. *Use Active Voice*

Don't say: "The paper was placed on the wall by the doctor."

Use active voice: "The doctor placed the paper on the wall."

6. *Avoid Omniscient Narrator*

Books have more immediacy if you stay within one character's head. Using this approach, the narrator does not have knowledge of what other people are thinking. This is called "third-person limited perspective." For example, if we are "in" Jake's head, we are in Jake's head for most of the book. We can't suddenly know how Melinda is feeling. Jake doesn't read her mind. The novelist can suggest how she feels through Jake's opinions and what he sees and hears, and what she says and does. This rule is not firm, and sometimes authors switch point of view with the start of a chapter, but I think it is best to stay in the head of one character. Some people use an omniscient narrator who can describe things that the protagonist does not know, but many of the best books avoid it.

7. *Don't Rush the Scene*

If a scene sounds rushed, with too little attention to detail and texture, then more words are needed to draw out the action and suspense.

8. *Natural Dialog*

If you are unsure if the dialog sounds natural, read it out loud to yourself. This is a great way to make sure the dialog is realistic.

9. *Involve All Senses*

The reader will become immersed in the book if you stimulate the reader's senses. For example, if you've gone two pages without stimulating the reader (and character in the book) with an odor, tactile feeling, sound, or taste, the book will have less immediacy.

10. *Use "Said"*

Some beginning writers dislike using "said" too often and replace "said" with words like "commanded," "remarked," "uttered," "began," and so forth. Perhaps beginners feel that too many "saids" stick out. But nothing can be further from the truth. It is usually much worse to continually insert words like muttered, uttered, yelled, and the like. The best writers use "said" to specify who is talking, and they let the dialog convey the meaning. For example,

This is bad: "Get out of here now!" he commanded.

This is good: "Get out of here now!" he said.

The word "commanded" is an unnecessary distraction. In any case, it's obvious the sentence is a command. When readers read "said," their eyes barely pause. The "said" goes almost unnoticed. This is what you want. Replacement words, such as "remarked," stick out obtrusively, which is what you don't want. Some authors don't even use "he asked" for questions. Rather, they write: "Where is it?" he said.

11. *Don't "Begin to"*

Don't have your characters "begin to do something," "try to do something," "start to do something," and so forth. Just have them do it. Example: "Mary began to skip down the block" might be changed to, "Mary skipped down the block."

12. *Avoid Repeated Use of "As He"*

Avoid too many "as he" constructs. Example: "Mary turned on the TV as she looked at the clock." You could change this to: "Mary turned on the TV and looked at the clock."

13. *Provide Character Reactions*

Example: When something is said or done to a character that is out of the ordinary, have the character respond. New writers often forget to show the responses of characters before moving on with the plot.

14. *Which or That?*

Use "which" with a comma when the phrase can be set off with parentheses without changing the meaning of the main part of the sentence. Examples: (1) I like dogs that bark. (2) I like the German Shepherd species, which has pointed ears, a tan coat, and teeth that rip.

Additional suggestions for improving the mechanics of your writing are listed in Note 15 to this chapter.

Creating a Great Novel

Let's take a break from writing mechanics to give you a hot recipe for creating your first novel. In my opinion, if the author creates compelling characters and natural dialog, he or she is 90 percent of the way to success. The *way* in which a writer writes is actually more important than the plot. Stephen King, who writes so well, could write a novel about a peanut butter sandwich and it would be great. Here are some steps to consider when writing a novel that will delight readers:

1. Buy a *National Geographic* magazine. Page through it and select a setting for the novel. Look at the photos to help you create vivid descriptions.
2. Your novel should have two main characters $C1$ and $C2$ (a man and a woman) and two secondary characters $C3$ and $C4$ (also a man and a woman). $C1$ should fall in love with $C2$ during the course of the book, or, if already in love, their love should deepen. A subliminal attraction should also exist between $C1$ and $C4$ to increase tension. Character $C1$ should have a special skill that will help him (or her) solve problems presented in the book.
3. A dangerous condition should exist throughout much of the book.
4. The dangerous condition should appear to be mitigated at some point in the book but come back to haunt the characters.
5. Avoid any long descriptions that slow down the pace. Practice keeping the pace of your novel brisk by allowing no paragraph to be more than five sentences. (You can relax this prohibition later in a few places if you find it absolutely necessary.)
6. Start your book with something that grabs the reader's attention.
7. Make the first two sentences of the book shine.
8. Don't use flashbacks. They break the flow of the book.
9. Never submit a work to a publisher or agent without having several people proofread it first.

Even minor characters must have texture. Imagine your character is in a store buying a pack of cigarettes. Here is the bland version:

Cliff approached the cashier. "Could I have a pack of Marlboros?"

"Yeah," the cashier said.

The cashier is a total nothing, a shadow. Consider adding texture:

Cliff approached the cashier. "Could I have a pack of Marlboros?"

The cashier grinned and put on the tiniest glasses that Cliff had ever seen. "That stuff will kill you."

David Morrell writes in *Lessons from a Lifetime of Writing*, "Make it a habit to respect your walk-on characters and find efficient ways to make them seem fuller."

Marcel Proust asserted that readers will imagine the characters in books to be themselves and people similar to those the reader already knows. For example, the romantic partners in the book will take on characteristics of people whom readers have been drawn to in the past. "One cannot read a novel," Proust said, "without ascribing to the heroine the traits of the one we love." If you have characters in your novels that eventually part ways, recall Proust's observation about real life: "When two people part, it is the one who is not in love who makes the tender speeches."

Two outstanding guidebooks will help you polish and sell your books. Novelists should buy David Morrell's *Lessons from a Lifetime of Writing* to learn how to write fiction. Both fiction and nonfiction writers should buy *Jeff Herman's Guide to Book Publishers, Editors, & Literary Agents* to learn how to get published. Jeff Herman's *Guide* lists all the editors' names and interests.

If you think having a formula or recipe for creating a bestseller is farfetched, you should read Michael Maxen's article, "How to Manufacture a Best Seller."[16] In the article, Maxen gives ten steps. Here are just a few:

- The hero is an expert.
- The villain is an expert.
- The hero has a team of experts in various fields working behind him.
- Two or more on the team must fall in love.
- Two or more on the team must die.
- All deaths must proceed from the individual to the group. My example: never start with an earthquake that kills 30,000 people. Instead, start with "Mike was walking down Fifth Avenue when a lamppost began to tilt, and the birds screeched."

The Lester Dent Master Fiction Plot

Lester Dent (1905–1959) was author of over a hundred Doc Savage pulp stories written under the name of Kenneth Robeson. He also wrote numerous Western, air-war, detective, and mystery stories. Dent attributed his success to his marvelous recipe for plotting a best seller and said, "No yarn of mine written to the formula has yet failed to sell."

Dent likened the creation of a story to constructing a building. Let me paraphrase his basic plot ideas, occasionally adding some of my own examples:[17]

1. The writer's first step is to determine a weapon for murder, e.g., gun, knife, cyanide, scorpion, genetically engineered virus, or hallucinogenic worm. The weapon will suggest details of the plot; for example, you will have to think about how scorpions kill. Where would someone get a scorpion? How should they be handled without getting hurt?

2. Determine a location for the story. As I said, I do this by looking at *National Geographic* or by setting novels in my home town. You can give your readers and your editors the very realistic illusion that you have visited these locals just by reading *National Geographic* and letting yourself enter the details of the photos in the magazine.

3. Ensure that a menace "hangs like a cloud" over the hero.

4. Determine what the villain wants, e.g., jewels, the Holy Grail, a formula for a drug, a computer chip, a secret code, or the love of a Slovenian woman.

5. In the first paragraph or first line of the story, introduce the hero and "swat him with a fistful of trouble." The author should "hint at a mystery, a menace or a problem to be solved—something the hero has to cope with."

6. The hero's action should place him in an actual physical conflict near the end of the first 1,500 words, at which point a complete surprise twist in the plot development takes place.

7. Whatever the menace is, Dent says, "Don't tell about it! Show how the thing looked. Again, this is one of the secrets of good writing; never tell the reader—show him. (He trembles, roving eyes, slackened jaw, and such.)"

8. Put a tiny surprise on each page, for example, a door creaking open by the wind, a strange scratching sound from the roof, a drop of blood, etc.

9. The hero must be buried in troubles and must free himself using his own "skill, training, or brawn."

Dean Koontz

I once heard a story about an English professor who told her class that a good short story needs five elements: a religious reference, a sexual reference, lurking danger, brevity, and a person with expertise in a field.

Heeding the professor's advice, the next day a student submitted a story to the professor. The entire story consisted of only two sentences: *"My God!" the zoologist said. "A leech is clinging to your left breast."*

Novelist Dean Koontz says that a best seller should require collaboration between a hero and a partner: "I started with that premise— that change is accomplished not so much on your own, but by finding another person who brings something to you. That person has something you need and you have something that he or she needs in terms of emotional or intellectual interaction. Together you can to some degree transform yourselves, your future, your destiny."[18]

We should listen to Koontz. At least seven of Koontz's novels rose to number one on the *New York Times* hardcover best-seller list (*Lightning, Midnight, Cold Fire, Hideaway, Dragon Tears, Intensity,* and *Sole Survivor*), making him one of only ten writers ever to have achieved that milestone. Eleven of his books have risen to the number-one position in paperback. Koontz's worldwide sales are over 200 million copies.

On a personal note, I was able to publish more than 30 books since 1990 because of "drive," not "talent." Dean Koontz says that "If I've learned one lesson in life, it is that you can have all the ability in the world and it means absolutely nothing without perseverance."[19]

The Absurd, Sad History of Destroyed Manuscripts

I recall the saddest writing experience I had, while working on my book the *Loom of God.* I lost a whole weekend's worth of work because, as the weekend closed, I wanted to back up the file to a diskette. I issued in DOS:

Copy a:loom.doc loom.doc
Instead of
Copy loom.doc a:loom.doc

Naturally, my erroneous command copied an old version of the book from the diskette on top of the new one, and I lost everything I created during the long weekend. I almost cried.

However, this setback was nothing compared to some historical examples. According to Robert Hendrickson in *The Literary Life*, British historian Thomas Carlyle (1795–1881) had to rewrite the entire first volume of his *History of the French Revolution* when a maid burned the manuscript, mistaking it for trash. In a fit of rage, the wife of William Ainsworth (1805–1882) tossed the manuscript for his Latin dictionary into the fire, and it took Ainsworth three years to rewrite it.

I usually completely finish writing a book before I attempt to get a publishing contract for the book. This means that once the contract is signed, the book is essentially ready to be delivered. However, usually a year to 18 months passes between the time I send the final manuscript to the publisher and the book is actually published. This delay reminds me of Harold Brodkey's novel *Runaway Soul,* which was contracted for in 1960 and not published until 31 years later! The delay was due to Brodkey's slowness and changes in publishers. During this period, perhaps he was the most famous "author" to become famous for not writing a book.

My Editors, Shirley Jackson, John Cheever

I have had many editors over the course of my book publishing career. All have been kind and friendly, although I am always sad when an editor is fired or leaves the publisher just as the book is contracted for or published. That's happened a few times. As a result, the book usually gets less attention, and sales decline. Sometimes editors have asked me to cut material to decrease the size of the book, but we've never had a quarrel. I've met some of my editors in person, and others I've "met" only through e-mail or by phone.

I once had a minor disappointment with my publisher Wiley, who charged me a $1,000 fee for creating an index to my book *Black Holes: A Traveler's Guide*. Nothing in the contract gave them permission to do this, and I always do my own indexes, which take me just a few hours. Alas, although I brought this to my editor's attention, I never did get my $1,000 returned.

Perhaps one of the strangest quarrels between an editor and publisher involved Shirley Jackson (1919–1965) and her publisher Alfred Knopf. Jackson is the prolific American writer best known for her haunting short story "The Lottery." She also dabbled in magic and voodoo. I never did determine what her quarrel was about, but she is said to have made a wax image of her editor and stuck a pin in one leg. Knopf then broke his leg skiing.

For the most part, an editor does not edit your book—at least this is my personal experience, mostly with nonfiction books. An editor's main goal, so far as I can tell, is to identify books that will sell and bring some prestige to a publisher. Editors show book proposals to their marketing people, and if the marketing people decide the book won't sell, the editor is overruled.

I have had one or two editors that carefully read my books and offered very detailed suggestions for improvements. However, for the most part, actual editing is done by hired copyeditors who check for grammar and other simple problems and generally try to polish a book so that it shines. Many of the copyeditors I have worked with have been excellent. My university publishers—like Oxford, Cambridge, and Princeton University Press—always send the book to outside technical reviewers who provide detailed commentary and suggestions.

Most of the translators for foreign editions of my books have never contacted me, but some translators have been very involved with me for my math books and have caught errors and approached me with questions.

Sometimes I don't even know when a foreign edition of my book is published, and I find new foreign covers simply by typing "Pickover" into the Google image search. When foreign sales are negotiated by the publisher, I split the advance 50-50 with the publishers. My books have been published in French, Greek, Italian, German, Japanese, Portuguese, Chinese, Korean, Spanish, Turkish, Greek, and Polish.

What is the definition of a good editor? American novelist John Cheever (1912–1982) once said, "My definition of a good editor is a man who I think charming, who sends me large checks, praises my work, my physical beauty, and my sexual prowess, and who has a stranglehold on the publisher and the bank."[20] This kind of editor is uncommon, and, as I've said, I've found that even if an editor really wants to publish one of my books, the marketing side of the publishing house can tell the editor, "no."

Book Promotions and the Mystery of Gertrude Stein

My book *The Zen of Magic Squares, Circles and Stars* was my only one in which the publisher used a physical object, namely a plastic sliding puzzle embossed with the name of the book, to help promote the book. Simon & Schuster used a similar gimmick when promoting their children's book *Doctor Dan and the Bandage Man*. The publisher decided to give away several Band-Aids with each copy. The publisher wired a friend at Johnson & Johnson: "Please ship half-million Band-Aids immediately." The friend replied, "Band-Aids on the way. What the hell happened to you?"[21]

I've done numerous book signings to promote my books. The signings were often preceded by a lecture. For the most part, book signings are a waste of time, because, unless you're Anne Rice or Stephen King, your attendance will be small. Bookstore signings do have some value in that they push the books into the store, and the books usually remain in the store for some time to encourage sales. My best book signings, in terms of books sold, were at the Smithsonian Institution in Washington, D.C., and at the 92nd Street YMHA in New York City. Sadly, even though my audience was huge at the YMHA, a fire alarm rang just as my lecture ended, which severely decreased book sales at the event.

Avant-garde American writer Gertrude Stein (1874–1946) began her now-legendary American lecture tour in 1934. It amazes me that perhaps no one, maybe not even the author herself, knew what Stein's writing was all about. "I wonder if you know what I mean," she mused to her audience on one occasion. "I do not quite know whether I do myself." Yet Stein was such a celebrity that 15 reporters sailed out to meet her ship in New York harbor.

The work that Stein always considered her masterpiece, *The Making of Americans*, published in Paris in 1925, sold about one hundred copies. It consists of 904 pages of long, repetitive, stream-of-consciousness sentences such as this: "Soon then there will be a history of every kind of men and women and of all the mixtures in them, sometime there will be a history of every man and every woman who ever were or are or will be living...." Stein seemed to reject punctuation, which she said was "necessary only for the feeble-minded."

For a long time, I had wondered how Stein got along with her editors and publishers. What did they think of her work? I finally discovered that publisher Bennett Cerf published the following note on the dust jacket of Getrude Stein's *The Geographical History of America or the Relation of Human Nature to the Human Mind,* which was one of her 26 books:

> I do not know what Miss Stein is talking about. I do not even understand the title. I admire Miss Stein tremendously, and I like to publish her books, although most of the time I do not know what she is driving at. That, Miss Stein tells me, is because I am dumb.[22]

Book Dedications

I've experimented with many unusual "dedications" for my books over the years. One of my favorite dedications was for *Keys to Infinity*, which read: *This book is dedicated to all those who do not have a book dedicated to them.*

I dedicated my book *Wonders of Numbers* not to a person but to the Apocalyptic Magic Square in which all of its entries are prime numbers (divisible only by themselves and 1), and each row, column, and diagonal sum to 666, the Number of the Beast:

3	107	5	131	109	311
7	331	193	11	83	41
103	53	71	89	151	199
113	61	97	197	167	31
367	13	173	59	17	37
73	101	127	179	139	47

The Apocalyptic Magic Square

Many of my books have various codes in their dedications. Look at this strange dedication to my book *The Alien IQ Test*:

> To the chordates, to Amphioxus
> This book is also dedicated to those:

23815 251295225 208525 81225 25514 1242132054, 1144 2015
20815195 23815 81225 25514 1242132054. 1121915
2015 1291292021, 129129208, 1211391, 1144
2085 131185—2085 891920151893112
1618545351919151819 156 2085
82516141715793 "15124 817"
16851415135141514.

No one has ever decoded this message, although I can tell you that amphioxus and its kin are aquatic creatures that scientists feel are the evolutionary gateway to humanity, to more developed brains, and to consciousness. Amphioxus, the gate, born millions of year ago…Why do humans exist? The answer quite simply is that amphioxus and its close relatives were born eons ego and lived to survive.

My favorite dedication is one by poet and novelist e. e. cummings, who had to self-publish a book of poetry in 1935, because all the big-name publishers had rejected it. His dedication reads:

No thanks to: Farrar & Rinehart, Simon & Schuster, Coward-McCann, Limited Editions, Harcourt, Brace, Random House, Equinox Press, Smith & Haas, Viking Press, Knopf, Dutton, Harper's Scribner's, Covivi, Friede.[23]

All of these publishers had rejected his manuscript.

Bibliomania

I love collecting books, and occasionally face the sad task of discarding older books in order to make room for new ones. I do not know if many psychiatrists consider "bibliomania" to be an obsessive-compulsive disorder, but the lengths to which some individuals go to handle, possess, and accumulate books seems to be a form of madness. Some bibliomaniacs purchase so many books that their social relations and health are jeopardized.

In the 1700s, British bibliophile Thomas Rawlinson (1690–1755) crammed his room at Gray's Inn so full of books that he had to sleep in the foyer. He finally moved into a large mansion and filled it to over-flowing. When he died at age 44, there was essentially no room to sit in the house.

Monsieur Boulard, an 18th-century Parisian bibliomaniac, indiscriminately bought 600,000 books. They were stacked in cupboards, attics, and cellars. Alas, the books were so heavy that the house began to collapse. Boulard bought six more houses and filled them entirely with books.

British book collector Richard Heber (1774–1833) wanted to own multiple copies of every book ever published. He believed that every gentleman needed at least three copies of a book, one for his own reading, one to lend to friends, and one for his country house. In the end, he probably owned about 170,000 books stored in several houses. Holbrook

Jackson (1874–1948), English journalist, editor, author, and fellow bibliomaniac described Heber:

> A bibliomaniac if ever there was one. A bibliomaniac in the most unpleasant sense of the word; no confirmed drunkard, no incurable opium-eater had less self-control; to see a book was to desire it, to desire it was to possess it; the great and strong passion of his life was to amass such a library as no individual before him had ever amassed. His collection was omnigenous, and he never ceased to accumulate books of all kinds, buying them by all methods, in all places, at all times.[24]

Upon the death of Heber, British author and librarian Thomas Dibdin (1776–1847) was the first to break into Heber's home and view its contents. He writes,

> I looked around me in amazement. I had never seen rooms, cupboards, passages, and corridors, so choked, so suffocated with books. Treble rows were here, double rows were there. Hundreds of slim quartos— several upon each other—were longitudinally placed over thin and stunted duodecimos, reaching from one extremity of a shelf to another. Up to the very ceiling the piles of volumes extended; while the floor was strewn with them, in loose and numerous heaps. When I looked on all this, and thought what might be at [his other homes] it was difficult to describe my emotions.[25]

Officials soon found three houses in England crammed with Heber's books. Other equally stuffed houses were located in Antwerp, Ghent, Paris, and Brussels. No one knows if he had other hidden warehouses of books scattered around Europe.

Finally, one of the greatest bibliomaniacs of all time was the British bibliophile Sir Thomas Phillipps (1792–1872). He lived a miserly life in order to devote his life to his obsession. He gradually acquired 100,000 books and 60,000 manuscripts, which at the time was more than all the libraries of Cambridge University held. Phillipps purchased some of his manuscripts at a sale of Heber's manuscripts in 1836.

Phillipps was an individual with symptoms of classic obsessive-compulsive disorder. He never threw away a scrap of paper, hoarding household bills and copies of correspondences. Although he collected priceless manuscripts, he also acquired old records thrown out by the government, and cartloads of wastepaper on the way to be pulped.

In order to catalog his ever-growing collection, Phillipps enlisted the help of his wife and three daughters, who worked like slaves to list manuscripts. When his dining room became filled to overflowing, Phillipps simply locked it, leaving his wife and three daughters to make do with three poorly furnished bedrooms and a sitting room. Lady Phillipps's dressing-table was the only area in the bedroom that remained free of books. After a few years of living like a rat trapped in a maze, she could no longer endure the never-ending inflow of books. She became a drug addict as the only means of escape, and died at the age of 37.

After the death of his wife, Phillipps searched for a replacement wife with more money to help him finance his collection obsession. He eventually settled for a clergyman's daughter and collected with renewed vigor. "I wish to have one copy of every book in the world!!!!!" he wrote to a friend. He ate and slept among his books. His new wife eventually complained of rats, and had a nervous breakdown. Phillipps only watched as she was carted away to a cheap boarding house.

Phillipps died at the age of 80. Sifting through his books took several generations. Sales of his library's valuable contents continued into the 20th century as great treasures were gradually discovered.

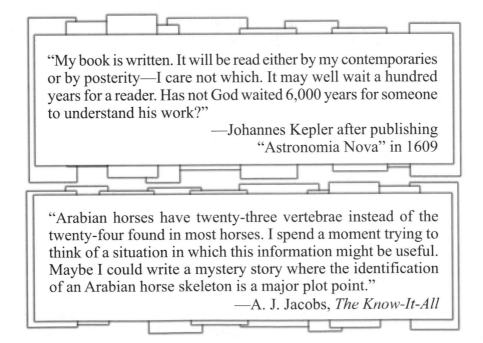

"My book is written. It will be read either by my contemporaries or by posterity—I care not which. It may well wait a hundred years for a reader. Has not God waited 6,000 years for someone to understand his work?"

—Johannes Kepler after publishing
"Astronomia Nova" in 1609

"Arabian horses have twenty-three vertebrae instead of the twenty-four found in most horses. I spend a moment trying to think of a situation in which this information might be useful. Maybe I could write a mystery story where the identification of an Arabian horse skeleton is a major plot point."

—A. J. Jacobs, *The Know-It-All*

"Most well-educated people have not finished Proust's *A la recherche* [*In Search of Lost Time*] for the same reason that most well-traveled people have not visited Antarctica: it is a long and expensive journey of uncertain value."

—Daniel Mark Epstein,
"Proust regained," *The New Criterion*

"The United Nations defines a book as a text that is at least forty-nine pages long."

—A. J. Jacobs, *The Know-It-All*

"It was so easy to write my novel, but how difficult it will be to publish it!"

—Marcel Proust, in George Painter's
Marcel Proust: A Biography

"I do not know personally ten people who have read Proust's novel whole, and yet the work cannot really be appreciated any other way. Like a cathedral, it must be seen from all points of the compass, and from inside and out, to understand its perfections."

—Daniel Mark Epstein,
"Proust Regained," *The New Criterion*

"Marcel Proust may have had a hard time remembering things past, considering his fondness for opium, morphine, hypnotics, camphor cigarettes, and possibly heroin, not to mention booze."
—Russ Kick, *The Disinformation Book of Lists*

"To form an insane obsession over a defiant and difficult lover is necessary for the production of superb literature. This is certainly what Proust is all about, and why *Remembrance of Things Past* is the greatest novel ever written."
—Charles Mudede, "What Sickness Is Good For: The Greatest Novel Ever Written," TheStranger.Com

"I told my friends of [Proust's] daft book that took half an hour to describe someone ringing a doorbell. Inside, however, my dismissive tone was tempered. Secretly, I was impressed."
—Stephen Mitchelmore, "Perchance To Dream"

Neoreality and the Quest for Transcendence

In which we encounter Einstein, Rumi, God, the anthropic principle, Stephen Hawking, the Bible, Proust's hyper-realities, The Lobotomy Club, Sushi Never Sleeps, *fractals, brain surgery, Italian filmmaking, neorealism, stellar nucleosynthesis, the Big Bang, Paul Davies, Frank Tipler, Marcel Proust, H. P. Lovecraft, Andrei Linde, Sir Fred Hoyle, Rudy Rucker, Robert Jastrow, The Templeton Foundation, multiple universes, Paul Kammerer, synchronicity, and the shoreless sea of love.*

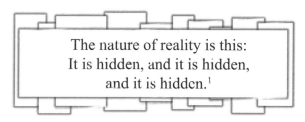

The nature of reality is this:
It is hidden, and it is hidden,
and it is hidden.[1]

Mysterium Tremendum

As we've discussed, Marcel Proust sometimes seemed to live in a fractal hyperreality, a universe filled with a cascade of details and observations beyond our ordinary ken. While walking in the town of Combray, Proust takes us on a journey along Guermates way, when suddenly reality awakens and speaks to him: "A rooftop, a reflection of sunlight on stone, the smell of a path would make me halt because of the singular pleasure they were giving me...and also because they seemed to be hiding something that invited one to come take."[2] Proust contemplated the exact line of the roof, the colors of the stone—simple objects that seemed just on the threshold of delivering something secret, as if they were a facade for the fabric of reality that lay only millimeters beyond. He stood on the threshold of something that transcended space and time, a *mysterium tremendum.*

Proust believed that reality and our perception of reality often were indistinguishable: "There is no great difference between the memory of a dream and the memory of a reality."[3]

Thinking about Proust's strange realities, I developed several novels that dealt with what I call "neorealities." Charles L. Harper, Jr., Executive Director and Senior Vice President of the Templeton Foundation, was so intrigued with my ideas that he asked me to expound on them in a book he edited titled *Spiritual Information*.[4] The John Templeton Foundation was established in 1987 by renowned international investor Sir John Templeton to act as a catalyst for ideas at the boundaries between science and religion.[5] In the following pages I cover many of the topics discussed in my work for this foundation.

Neoreality

What is reality? What is transcendence? How can we open our minds so that we can reason beyond the limits of our intuition? When Albert Einstein was asked about reality, he replied, "Reality is merely an illusion, albeit a very persistent one." In an effort to stretch readers' minds, I have considered both Einstein and Sufi poet Rumi while publishing thirty books on topics on the borderlands of science and religion.

Very recently, and perhaps most importantly, I published four science-fiction novels in a "Neoreality" series in which both the reader and protagonists cope with realities separated from ours by thin veils. These distortions and parallel universes are a backdrop for human emotion, scientific logic, grand adventure, and a variety of religious discussions. For example, in *Liquid Earth*, religious robots help humans cope with a reality that melts along a rustic Main Street in Shrub Oak, New York. In *The Lobotomy Club*, a group of people perform brain surgery on themselves in order to see religious visions and a "truer" reality. In *Sushi Never Sleeps*, readers ponder a fractal society with inhabitants living at different size scales. Would different population groups, because of size, develop their own separate societies, religions, and laws? Would some of the tiny Fractalians believe that individuals a million times their size even exist, or would the humongous, uliginose beings be relegated to the realm of mythological creatures, like the superhuman gods of yore? And, finally, in *Egg Drop Soup*, an alien object allows people to explore countless realities populated by a host of mysterious beings.

When we discussed H. P. Lovecraft's mysterious story "From Beyond" in Chapter 1, it was clear that he blended fantastic, dreamlike logic with ordinary life. Lovecraft explored the mystery and danger that occurs between our world and a world beyond. His characters were often reclusive bookish men—a bit like me—who tripped into new universes resembling a dark version of those in my Neoreality series. According to science-fiction writer Fritz Leiber Jr. (1910–1992), Lovecraft was one of the first "fantasy" writers who "firmly attached the emotion of spectral dread to such concepts as outer space, the rim of the cosmos, alien beings, unsuspected dimensions, and the conceivable universe lying outside our own spacetime continuum."[6] Although Lovecraft, like me, received many of his ideas in dreams and also tended to be a skeptic, many fans have postulated that he actually had direct unconscious experiences of these other realms.[7]

Sometimes readers of the Neoreality series ask me why I write on God, strange realities, and religious subjects. I tend to be skeptical about the paranormal. However, I do feel that there are facets of the universe we can never understand, just as a monkey can never understand calculus, black holes, symbolic logic, and poetry. There are thoughts we can never think, visions we can only glimpse. It is at this filmy, veiled interface between human reality and a reality beyond that we may find the numinous, which some may liken to God.

But what exactly is *neoreality*? In my Neoreality book series, I use the word "neoreality" to imply a new or altered reality that so closely resembles ours that the differences are usually imperceptible. These realities are often futuristic, fresh, and alive with detail. Readers find themselves in touch with a hyperreality, a religious reality beyond space and time. Odd portals help characters transcend ordinary existence. The word "neorealism" has *traditionally* described a movement in Italian filmmaking, characterized by the depictions of poor people and their daily challenges. In neorealistic movies, directors often featured ordinary characters in plots that meandered like wisps and eddies of wind. The directors did a minimum of editing and fancy camerawork. My new use of the word neoreality is not synonymous with neorealism, although I can resonate with the old neorealistic characters, buffeted by the seemingly random circumstances around them. Navigating the chaotic churn, and speculating about God, is the very essence of adventure.

God and the Anthropic Principle

Belief in an omniscient God and the promise of heaven are important ideas to adherents of great monotheistic religions such as Christianity, Judaism, and Islam. These beliefs pervade much of Western culture and are clearly evident in the U.S. Recent surveys indicate:

- ◆ The U.S. ranks highest, along with Iceland and Philippines, for those who believe in heaven (63% of the U.S. population).
- ◆ The U.S. ranks highest for those who believe in life after death (55%).
- ◆ 84% of Americans believe that God performed miracles.[8]

Indeed, science and religion are *both* thriving in America.

Not only do many *lay people* believe in God, but various *scientists* have used evidence from physics and astronomy to conclude that God exists. Note, however, that the scientists' "God" may not be the God of the Israelites, who smites the wicked, but rather a God that established various mathematical and physical parameters that permitted life to evolve in the universe. Some scientists believe we exist because of cosmic coincidences, or more accurately, we exist because of seemingly "finely tuned" numerical constants that permit life. Those individuals who hold to this *anthropic principle* suggest these numbers to be near miracles that might suggest an intelligent design to the universe. Here are just a few examples of where religion and science gently kiss.

We owe our very lives to the element carbon, which was first manufactured in stars before the Earth formed. The challenge in creating carbon is getting two helium nuclei in stars to stick together until they are struck by a third.[9] It turns out that this is accomplished only because of internal resonances, or energy levels, of carbon and oxygen nuclei. If the carbon resonance level were only four percent lower, carbon atoms wouldn't form. Were the oxygen resonance level only half a percent higher, almost all the carbon would disappear as it combined with helium to form oxygen.[10] This means that human existence depends on the fine-tuning of these two nuclear resonances. The famous astronomer Sir Fred Hoyle said that his atheism was shaken by facts such as these:

> If you wanted to produce carbon and oxygen in roughly equal quantities by stellar nucleosynthesis, these are just the two levels you have to fix. Your fixing would have to be just about where these levels are actually found to be… A common sense interpretation of the facts suggest that a superintellect has monkeyed with physics, as well as with chemistry and biology, and there are no blind forces worth speaking about in nature. The numbers one calculates from the facts seem to me so overwhelming as to put this conclusion almost beyond question…Rather than accept that fantastically small probability of life having arisen through the blind forces of nature, it

seemed better to suppose that the origin of life was a deliberate intellectual act.[11]

Robert Jastrow, the head of NASA's Goddard Institute for Space Studies, called this the most powerful evidence for the existence of God ever to come out of science.[12] Other amazing parameters abound.[13] If all of the stars in the universe were heavier than three solar masses, they would live for only about 500 million years, and life would not have time to evolve beyond primitive bacteria. Stephen Hawking has estimated that if the rate of the universe's expansion one second after the Big Bang had been smaller by even one part in a hundred thousand million million, the universe would have recollapsed.[14] The universe must live for billions of years to permit time for intelligent life to evolve. On the other hand, the universe might have expanded so rapidly that protons and electrons never united to make hydrogen atoms.

Paul Davies has calculated that the odds against the initial conditions being suitable for later star formation as 1 followed by a thousand billion billion zeros.[15] Paul Davies, John Barrow, and Frank Tipler estimated that a change in the strength of gravity or of the weak force by only one part in 10^{100} would have prevented advanced life forms from evolving.[16] There is no a priori physical reason these constants and quantities should possess the values they do. This has led the one-time agnostic physicist Paul Davies to write, "Through my scientific work I have come to believe more and more strongly that the physical universe is put together with an ingenuity so astonishing that I cannot accept it merely as a brute fact."[17] Of course, these conclusions are controversial, and an infinite number of random (non-designed) universes could exist, ours being just one that permits carbon-based life. Some researchers have even speculated that child universes are constantly budding off from parent universes and that the child universe inherits a set of physical laws similar to the parent, a process reminiscent of evolution of biological characteristics of life on Earth.[18]

Create Your Own Universe

We can go even further and think about the wild implications for multiple universes—such as those presented in my Neoreality series— and what they say about our power in relation to God's. Stanford University physics professor Andrei Linde has speculated that it might be possible to create a new baby universe in a laboratory by violently compressing matter at high temperatures—curiously, one milligram of matter may initiate an eternal self-reproducing universe.[19] What would be the economic or spiritual gain we would get from creating a universe, considering it would be extremely difficult, if not impossible, to enter the new universe from ours? Would God care if we created such universes at will? Andre Linde and writer Rudy Rucker have discussed methods for encoding a message for the new universe's potential inhabitants by manipulating parameters of physics, such as the masses and charges of particles, although this would be a precarious experiment given the difficulty of manipulating these constants so that they code a message and permit life to evolve. In light of the possibility of multiple universes, perhaps the term "omniscient" takes on a new meaning, and the God of the Old Testament might be omniscient only in the sense that He knows all that can be known about a single universe and not all universes.

God and Mathematics

Is God a mathematician? Certainly the world, the universe, and nature can be reliably understood using mathematics. Nature *is* mathematics. The arrangement of seeds in a sunflower can be understood using Fibonacci numbers. The shape assumed by a delicate spiderweb suspended from fixed points, or the cross-section of sails bellying in the wind, is a catenary—a simple curve defined by a simple formula. Seashells, animal horns, and the cochlea of the ear are logarithmic spirals, which can be generated using a mathematical constant known as the golden ratio. Mountains and the branching patterns of blood vessels and plants are fractals, a class of shapes that exhibit similar structures at different magnifications. Einstein's $E = mc^2$ defines the fundamental relationship between energy and matter. And a few simple constants—the gravitational constant, Planck's constant, and the speed of light—control the destiny of the universe. I do not know if God is a mathematician, but mathematics is the loom upon which God weaves the fabric of the universe.[20]

I think that our brains are wired with a desire for religion and belief in an omniscient God. If this is so, the reasons for our interest, and the rituals we use, are buried deep in the essence of our nature. Religion is at the edge of the known and the unknown, poised on the fractal boundaries of history, philosophy, psychology, biology, and many other scientific disciplines. Because of this, religion and religious paradoxes are an important topic for contemplation and study. Even with the great scientific strides we will make in this century, we will nevertheless continue to swim in a sea of mystery. Humans need to make sense of the world and will surely continue to use both logic and religion for that task. What patterns and connections will we see in the 21st century? Who and what will be our God?

Reality, Waves, Love

Austrian biologist Paul Kammerer once compared events in our world to the tops of waves in an ocean. We notice the tops of the isolated waves, but beneath the surface there may be some kind of synchronistic mechanism that connects them. Whatever you believe about such far-out speculation, be humble. Our brains, which evolved to make us run from lions on the Ethiopian plains, may not be constructed to penetrate the infinite veil of reality. We may need science, computers, brain augmentation, and even literature and poetry to help us tear away the veils. Einstein himself realized the insufficiency of the human mind when he wrote, "My feeling is religious insofar as I am imbued with the consciousness of the insufficiency of the human mind to understand more deeply the harmony of the Universe which we try to formulate as 'laws of nature.'"[21] For those of you who read the Neoreality book series, look for the hidden mechanism, feel the connections, pierce the cosmic shroud, and sail on the shoreless sea of love.[22]

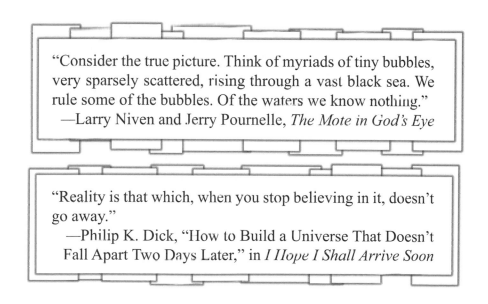

"Consider the true picture. Think of myriads of tiny bubbles, very sparsely scattered, rising through a vast black sea. We rule some of the bubbles. Of the waters we know nothing."
—Larry Niven and Jerry Pournelle, *The Mote in God's Eye*

"Reality is that which, when you stop believing in it, doesn't go away."
—Philip K. Dick, "How to Build a Universe That Doesn't Fall Apart Two Days Later," in *I Hope I Shall Arrive Soon*

"There was, in Proust's very physical appearance, in the atmosphere that he created about himself, something so remarkable that, seeing him, one had a feeling akin to amazement. He did not belong to the common run of mankind, but always produced the impression that he was a figure of a nightmare, or of a different age, almost of a different world— but of what world?"

—Edmond Jaloux, 1917,
quoted in André Maurois's *Proust: A Biography*

"The various forms of reality are akin to viewing life through different perspectives. Solutions to problems may rear their beauteous heads when viewed through the correct lens. A personal trainer may use imagery that ranges from biophysical to mystical in order to induce a client to obtain the best way to practice an exercise set, and to provide sufficient motivation for the client to actually perform such an otherwise dreadful regimen."

—Jamie Forbes, personal communication

"Guerilla philosopher and reality hacker Robert Anton Wilson believes that reading aloud James Joyce's experimental novel *Finnegans Wake* triggers altered states."

—Russ Kick, *The Disinformation Book of Lists*

"God gave us the darkness so we could see the stars."

—Johnny Cash, in "Farmers Almanac"

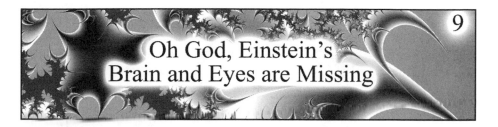

Oh God, Einstein's Brain and Eyes are Missing

In which we encounter Einstein's brain, Maja Einstein, the Sylvian fissure, Einstein's disembodied eyes, Dr. Henry Abrams, Einstein's disembodied hair, Einstein's pipe, the legend of Mochaoi, Henri Bergson, spacetime, time dilation, Marcel Proust, Thomas De Quincey, Angelus Silesius, Félirena Naomh Nerennachm, Philip K. Dick, Mount Coelian, the Seven Sleepers, God's time, Moslem legends, magical horse Burak, Isaac Newton, Hollywood, Marian Diamond, Stephen Hawking, the Chronology Projection Conjecture, psychoactive ketamine, the special theory of relativity, In Search of Lost Time, *and future-life progression. Is time travel possible? If time is something learned, can we unlearn it? Can the flow of time be stopped? What if Einstein had lived another twenty years?*

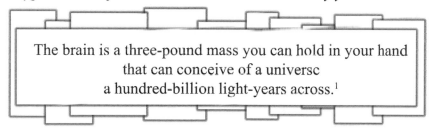

The brain is a three-pound mass you can hold in your hand
that can conceive of a universe
a hundred-billion light-years across.[1]

Where Is Einstein's Brain?

While in college, gossiping with other biology students in the cafeteria over root beers and burgers, I overheard mysterious, sordid tales about Einstein's brain. According to my friends, Einstein's big brain had been removed from his corpse without permission, and some mad doctor kept it in a Mason jar next to a beer cooler deep in the heart of Kansas. Could these rumors possibly have been true? I've learned quite a few new things since I first discussed this topic in my book *Strange Brains and Genius.*

Nobel Prize–winner Albert Einstein (1879–1955) is recognized as one of the greatest physicists of all time and the most important scientist of the 20th century. He proposed the special and general theories of relativity, which revolutionized our understanding of space and time. He also made major contributions to the science of quantum mechanics, statistical mechanics, and cosmology.

What might we learn if we could actually study the folds of Einstein's brain? People have always wondered if his brain or even his skull was different from ordinary people. His sister Maja Einstein noted that her mother was "shocked at the sight of the back of Albert Einstein's head, which was extremely large and angular, and she feared she had given birth to a deformed child." Astronomer Charles Nordmann in 1992 said: "Einstein's skull is clearly, and to an extraordinary degree, brachycephalic, great in breadth and receding towards the nape of the neck without exceeding the vertical."

What are we to make of all these odd musings? Although we don't have the brains of Marcel Proust, Kemmons Wilson, or other great individuals discussed in this book available to study, scientists today should be able to determine if Einstein's brain differs from the brains of mere mortals. His brain still exists. For years, however, his brain was missing. Just thinking about it gives me the same shivers I felt while watching *Night of the Living Dead*.

For years after Einstein's death, his brain was in the custody of Thomas S. Harvey, M.D., the pathologist who performed Einstein's autopsy in 1955—without permission.[2] Harvey had kept the brain in a jar behind a beer cooler. Earlier, Harvey hid the brain in the basement of his ex-wife, who had reportedly said, "I wish someone would get rid of the damn thing!"[3]

According to some journalists, Einstein's brain was saved at his son's request. Other sources suggested that Einstein himself wanted his brain saved, but Einstein's estate denies this. Contrary to what you might think, research on Einstein's brain did not take place until 30 years after his death—perhaps because no one could agree on how to cut up this priceless object. In 1978, *Science* magazine suggested that the delay was never satisfactorily explained.

The search for Einstein's brain began in 1976 when the magazine *New Jersey Monthly* assigned one of its reporters, Steven Levy, the task of finding the brain. Levy's detective work led him to Thomas Harvey's office in Wichita, Kansas, where Einstein's brain floated in a Mason jar packed in a cardboard box marked "Costa Cider." Levy reports that Harvey had little interest because "he had other things to do." In the August 1978 *New Jersey Monthly*, Levy tells his own feelings as he gazes at the chunks of Einstein floating in Harvey's jar:

I had risen up to look into the jar, but now I was sunk in my chair, speechless. My eyes were fixed upon that jar as I tried to comprehend that these pieces of gunk bobbing up and down had caused a revolution in physics and quite possibly changed the course of civilization. *There it was!*[4]

In 1981, *Science* magazine again asked Dr. Harvey if he had finally finished his research on Einstein's brain. Harvey's response was, "No concrete plan. I have my ideas about it but they have not solidified." *Science* magazine wrote, "Harvey possesses small fragments of the brain, but declines to say exactly where they are now stored. Einstein's estate, he says, has no interest in them."[5]

Finally, in 1985, neuroanatomists Marian Diamond and Arnold Scheibel persuaded Dr. Harvey to give them some tissue samples. Having already conducted experiments demonstrating that rat brains are larger when rats live in an intellectually stimulating environment, Diamond was curious to see if there was a similar effect on Einstein. After counting cells in Einstein's brain, they often found more glial cells than in normal brains. (Glial cells support the neurons, which are the functional units of the brain.) Diamond and colleagues studied "area 9" and "area 39," the two regions of Einstein's superior prefrontal and inferior parietal lobe from which cells were counted in both hemispheres. Diamond and Scheibel chose these areas because they were concerned with "higher" neural functions such as the capacity for abstraction, calculation, planning and establishment of behavioral strategies, attention, and imagery. Neurons in these areas don't receive primary sensory information but rather interpret impulses sent to them from other areas of the brain. According to the researchers, Einstein had more glial support cells because his neurons (the workhorses of the brain) had greater metabolic need as a result of his unusual conceptual powers.

Oddly enough, Diamond and Scheibel found Einstein to have *fewer* glial cells than expected in Area 39 of Einstein's *left* brain hemisphere. Could Einstein's low number of glial cells account for his slower (or unusual) speech development? We know that this area impacts speech because lesions in this area lead to dyslexia. As we discussed in Chapter 2, Einstein once mentioned that written and spoken words were not important when he formulated his theories. Imagery and emotion actually played a greater role. In addition, Einstein said he had trouble learning to speak when he was a child. He reminisced in 1954: "My parents were worried because I started to talk comparatively late, and they consulted the doctor because of it. I cannot tell how old I was at that time, but certainly not younger than three." On the other hand, his maternal grandmother Jette and sister Maja said that he talked very well at an early age, so no one today is quite sure what to make of the conflicting statements regarding his speaking ability.

When Einstein was nine years old, he was still reported to be behind in speech development, and later in life he acknowledged his own "poor memory of words." Mathematician Ernst Straus (1922–1983), Einstein's assistant at Princeton Institute for Advanced Study, remarked in 1979:

> Einstein said that when he was between two and three years old, he formed the ambition to talk in whole sentences. If somebody asked him a question and he had to answer, he would form a sentence in his mind and then try it out on himself, thinking that he was whispering it to himself. But, as you know, a child is not very good at whispering so he said it softly. Then if it sounded all right, he would say it again to the person who had questioned him. Therefore, he sounded, at least to his nursemaid, as if said everything twice, once softly and once loudly, and she called him "Der Depperte" which is Bavarian for "the dopey one."[6]

The Marian Diamond studies are not without controversy. For example, Dr. Lucy Rorke, a neuropathologist in Philadelphia, feels that the number of glial cells was in fact normal, but she was impressed that the brain showed none of the usual degenerative changes one sees in older people. Rorke said it "looks like the brain of a young person."[7] Terence Hines of Pace University has also pointed out that the Diamond work may have statistical flaws in its analysis of Einstein's brain, so the debate continues.

Today, most of Einstein's brain is sectioned for microscopic study. My sources tell me that only his cerebellum and a piece of his cerebral cortex remain intact as they drift in a formaldehyde-filled jar under a beer cooler in Dr. Harvey's office.

Some of my best friends believe that, in the future, we will be able to resurrect a human being if a well-preserved brain is available. Alas, most of Einstein's brain is in pieces or slices. If Einstein ever hoped that his whole brain would be preserved for future revival by advanced technology, he would likely be disappointed. However, I think Einstein would have been miffed that his brain had been removed at all, because it had been his personal wish that his body be cremated. In fact, in 1955, his ashes were scattered at a secret location, possibly the Hudson River or Delaware River.

After his body had been cremated without the brain, Einstein's family learned about the brain "theft" and agreed to let Dr. Harvey keep the organ, so long as he did not use it for commercial purposes. Luckily for science, in 1996 Dr. Harvey gave a piece of the brain to Dr. Sandra Witelson, a neuroscientist at McMaster University in Ontario, Canada. She found that the brain's inferior parietal lobe was 15 percent wider than normal on both sides. She also found that the Sylvian fissure did not extend as far as in normal people. The absence of the groove may have allowed more neurons in this area to interconnect and thus cooperate more easily than in normal people.[8] Other parts of Einstein's brain were smaller than average, so that the brain as a whole was within normal range.

Looking for Einstein's Shining, Disembodied Eyes

In the previous section, we discussed the tale of Albert Einstein's brain, which was kept for years in a jar by a beer cooler. By now, a lot of people have heard that his brain was missing. But I find that most colleagues don't know that Einstein's eyes were also snatched from his face and hidden for 40 years in a bank vault in New Jersey. The physician who took the eyes visits them two or three times a year. Would you like to hear a few more details?

While working on a BBC documentary on Einstein's brain, Kevin Hull, the producer, made a stunning discovery: Einstein's eyes were removed by his opthalmologist, Dr. Henry Abrams of Loveladies, New Jersey. Abrams reportedly took the eyes with the permission of the hospital administrator. The removal required twenty minutes using a scissors and forceps.

Abrams felt that, when he looked into Einstein's disembodied eyes, he was looking into "into the beauties and mysteries of the world." He also said that Einstein's eyes "were angelic: they gave the impression he knew everything in the world. His eyes were godlike."[9] Colleagues tell me that the eyes were put up for auction in 1994, and I have since lost track as to their whereabouts. I think the auction never took place. I do recall that Abrams expected they could bring as much as $5 million.

Where on Earth Are Einstein's Hair and Pipe?

A few strands of Einstein's hair are still owned by John Reznikoff, president of University Archives, an independent manuscript and autograph dealership in Westport, Connecticut. Reznikoff is an extreme hair aficionado and has acquired hair from over a hundred celebrities including: Abraham Lincoln, John F. Kennedy, Marilyn Monroe, Albert Einstein, Napoleon Bonaparte, Elvis Presley, King Charles I, and Charles Dickens. So as not to overindulge, Reznikoff limits himself to acquiring only four new samples a year. Some samples of Einstein's hair was recently sold to a researcher who expects to extract the distinguished professor's DNA. A hundred years from now, we'll be able to clone these people. We might have a hundred Einsteins working together on problems.

Einstein left his violin to his only surviving grandson, Bernhard Caesar. Einstein's friend Abraham Pais received Einstein's last (smoking) pipe. The pipe's head is made of clay. The stem is made of some kind of reed material. Pais also inherited the galley proof to Einstein's "Generalized Theory of Gravitation."

Abraham Pais was a theoretical physicist and scientific historian, who died in 2000 at the age of 81. Pais worked with luminaries such as Einstein, Oppenheimer, Dirac, and Feynman. His best-known work was a biography of Einstein, *Subtle Is the Lord: The Science and the Life of Albert Einstein*, which was published in 1982 and won the 1983 American Book Award. Now that Pais is gone, I have lost track of Einstein's pipe.

Proust, Einstein, and the Dissolution of Time

Marcel Proust and Albert Einstein had a lot in common, and Proust was aware of some of Einstein's observations on the plastic nature of time. Perhaps Einstein's theories showing how the flow of time could change depending on an observer's motion or even on the strength of gravity inspired artists like Proust to reassess the meaning of time. Whether or not Einstein directly influenced Proust, *In Search of Lost Time* is famous for time dilation, in which time slows to a molasses-like crawl at certain points in the plot. Proust scholar Roger Shattuck in *Proust's Way* notes that "Incidents collect in a series of great pools…[that] engulf the landscape and give the impression of near motionless."

In Proust's novel, the town of Balbec, with its slow pace of life and attractive landmarks, seems trapped in time, like an ancient ant trapped in amber. I like to think of *In Search of Lost Time* as a collection of chambers connected by tunnels in a dense ant colony. In one chamber is an endless party taking place at one point in the narrator's life—and if we escape through a tunnel, we are quickly whisked away to another chamber in which time congeals yet again.

Gabriel García Márquez, the Nobel Prize–winning author, frequently uses the theme of time congealing to create a feeling of mystery. In his short story "Monologue of Isabel Watching It Rain in Macondo," the narrator is frozen in a sacerdotal stasis:

> The notion of time, upset since the day before, disappeared completely. Then there was no Thursday. What should have been Thursday was a physical jellylike thing that could have been parted with the hands in order to look into Friday.

Do newborn babies have a sense of the passage of time? Philosophers like Kant thought so, and they believed that time was something we experience directly from birth, that it exists outside of us. Other philosophers believe that time is a construct of the human mind. For example, philosopher Henri Bergson treats time as something entirely derived from subjective experience. According to Bergson, an infant would not experience time directly, but rather have to learn how to experience it. If time is something learned, can we unlearn it? Bergson also suggested that time does not consist of linear, clocklike measures of fixed and unchangeable moments. His theories of time influenced Proust, who didn't think of time as discrete moments, but rather as a flowing together of moments so that one point in time was not readily distinguishable from another.

Sometimes time seems to disappear entirely from Proust's work. We spend hundreds of pages examining the nature and ideas of a character or a situation, while there is minimal flow of time. As Jonathan Wallace writes in "Proust's Ruined Mirror": "In Proust's novel, time is a river in which the characters swim; it *tends* to carry them downstream, but like fish, they occasionally reverse themselves and struggle against its flow."[10] Proust's greatest desire was to travel through time, to recapture the past with its lost memories and people. In some ways *In Search of Lost Time* resembles a chunk of spacetime that contains past, present, and future. In this chunk, the reader and Proust may explore the story as they would a hyperspace palace, wandering in time and space through rooms anchored in different epochs.

According to Anatoly Vasilievich Lunacharsky's *On Literature and Art* (1934), Proust spent half his life reminiscing about his life using impressionistic brush strokes:

> This monstrous reworking of the first half of Proust's life becomes the second half. Proust the writer is no longer living, he is writing. The music and light of the true wave of life have no importance for Proust in his later years. The important thing is this astonishing chewing the cud of the past which is going on in all his 77 stomachs and which constantly renews this past and seems to deepen its implications. [It was as if Proust were recalling events in his life while] in the gently rotating convolutions of a mild sedative…savoring it with the concentrated attention one accords the complex bouquet of a uniquely rich wine.[11]

When Einstein was once asked about "psychological time," he remarked, "When you spend two hours with a nice girl, you think it's only a minute. But when you sit on a hot stove for a minute, you think it's two hours."

Since Einstein's death, there has been increasing research on the psychology of time dilation. For example, sleep studies show that, during dreaming, time is dilated; during brief periods of external time, there can be long sequences of internal events. If I woke you after you dreamed for five minutes, you could tell me a lengthy sequence of events that appeared to have taken much longer than five minutes.

In Chapter 4 we discussed how Ayahuasca users often felt that they were freed from time. Past, present, and future lose their significance. In Chapter 2 we discussed how certain languages may shape or reflect different cultures' notions of time. Your psychological perception of time is, of course, affected by such things as medications, time of day, your level of happiness, external stimuli, and even the temperature. Hypnosis can also cause time dilation, as can cannabis and LSD. Additionally, heat appears to speed up the activity of a chemical timepiece in the brain. For example, fever can severely speed your perception of time, perhaps partly because it speeds chemical processes. Opium is notorious for its effect on time perception. The English writer Thomas De Quincey reported that under the influence of opium he seemed to live as much as one hundred years in a single night. Another Englishman, J. Redwood Anderson, took hashish and said, "Time was so immensely lengthened that it practically ceased to exist." This reminds me of Tennyson's Lotus Land, "where it was always afternoon" or Samuel R. Delany's *Dhalgren* in which seconds sometimes lasted for hours.

Even without drugs, people can learn to stare at the second-hand of a clock and perceive it to stick, slow down, and hover. This takes training, but some people can experience the hand to stop altogether for a while. Some psychologists propose that the observing mind, the entity that correlates and makes sense of information submitted to it by the brain, is temporarily absent during these time-sticking periods. The brain hardware is left unattended while the mind has gone elsewhere.

A person under hypnosis can judge time more accurately than he can when in a normal waking state. For example, if a hypnotized person is asked to awake after five minutes, he can judge this time interval more accurately than normal. This leads me to believe that unconscious perception of time can be more accurate than conscious perception. By this I mean that the brain can be "trained" to measure certain time intervals in a fashion that the conscious mind cannot be trusted to measure. If I make a buzzing sound and then nine seconds later flash a light in your eyes, and do this over and over, your brain will eventually show a conditioned reflex: After I make a buzzing sound, nine seconds later your brain wave will change. However, if you were asked to make a conscious estimate of the delay between sound and light, your guess would be much less accurate.

The psychoactive compound ketamine can cause profound time distortions. According to writer James Kent, when the users enter a dissociative or "emergent" state, time slows to a "shuddering, thugging crawl—each moment stretching out into a sea of infinity and rolls sluggishly into the next. Seconds become minutes, minutes become hours, and eventually, in the peak, time ceases to have any meaning whatsoever. All branches of linear time bend in on themselves and collapse into a timeless and eternal state of simultaneous existence."[12]

What Is Time?

What is time? Is time travel possible?[13] For centuries, these questions have intrigued mystics, philosophers, and scientists such as Albert Einstein and Stephen Hawking. Much of ancient Greek philosophy was concerned with understanding the concept of eternity, and the subject of time is central to all the world's religions and cultures. Can the flow of time be stopped? Certainly some mystics thought so. Angelus Silesius (1624–1677), a German mystic of the Counter-Reformation, thought the flow of time could be suspended by mental powers. He wrote: "Time is of your own making; its clock ticks in your head. The moment you stop thought time too stops dead."

Today, physicists would agree that time is one of the strangest properties of our universe. There is a story circulating among scientists of an immigrant arriving in Manhattan who has lost his watch. He walks up to a man on Broadway and asks, "Please, Sir, what is time?" The scientist replies, "I'm sorry, you'll have to ask a philosopher. I'm just a physicist."

Most languages have a grammar with past and future tenses, and also demarcations such as seconds and minutes, and yesterday and tomorrow. Yet we cannot say exactly what time is. Although the study of time became scientific during the time of Galileo and Newton, a comprehensive explanation was given only in the 20th century by Einstein, who declared, in effect, that time is simply what a clock reads. The clock can be the rotation of a gear, the movement of the stars, a heartbeat, or vibrations of a cesium atom. A typical grandfather clock follows the simple Newtonian law that states that the velocity of a body not subject to external forces remains constant. This means that clock hands travel equal distances in equal times. Although this kind of clock is useful for everyday life, modern science finds that time can be warped in various ways, like dough in the hands of a cosmic baker.

Félirena Naomh Nerennachm and Philip K. Dick

Ancient legends of time distortion are quite common. In the *Félirena Naomh Nerennachm*, an early "Chronicle of Irish Saints," we encounter the legend of Mochaoi, Abbot of n'-Aondruim in Uladh (North Ireland). While in the woods, Mochaoi hears a beautiful bright bird singing in the trees. When he asks the bird why it is singing, the bird replies that it is singing to an angel. After a while, the bird goes to sleep, but Mochaoi continues to hear the angelic song for hundreds of years. Mochaoi doesn't age, and time stops in the woods around him. Only when he returns to his Church does he realize that centuries have passed. A shrine is built in Mochaoi's woods where angels congregate, "their wings flapping, their bodies trembling, waiting to send tidings of prayer and repentance with a beat of their wings to the King of the Everlasting."

Proust also seems to stop time with a magical bird:

Not a footstep was to be heard on any of the paths. Somewhere in one of the tall trees…an invisible bird, desperately attempting to make the day seem shorter, was exploring with a long, continuous note the solitude that pressed it on every side, but it received at once so unanimous an answer, so powerful a repercussion of silence and of immobility that, one would have said, it had arrested for all eternity the moment which it had been trying to make pass more quickly.[14]

Many modern novelists, philosophers, madmen, and provocateurs deeply believe that time is not what we think it to be. Novelist Philip K. Dick, for example, suggested that time on Earth has stopped in the year 50 AD, and he gives concrete reasons for his theory in his breathtaking essay "How to Build a Universe That Doesn't Fall Apart Two Days Later," in *I Hope I Shall Arrive Soon*. In short, he believes that our world today is not taking place in the 21st century, and we are deceived and live in a counterfeit reality lodged in a spacetime pocket in 50 AD. He writes:

My theory is this: *time is not real*.…Despite all the change we see, a specific permanent landscape underlies the world of change: and this invisible underlying landscape is that of the Bible; it, specifically, is the period immediately following the death and resurrection of Christ; it is, in other words, the time period of the Book of Acts.…[There is] internal evidence…that another reality, an unchanging one, exactly as Parmenides and Plato suspected, underlies the visible phenomenal world of change…and we can cut through to it.…Thousands of years pass, but the world of the Bible *is concealed beneath it*, still there and still real.

To Dick, the Bible is a literally real but veiled landscape, never changing but usually hidden from our sight. Dick cites numerous coincidences in his life that plunged him back to the time period of the Book of Acts. His novels contained surprising fragments of the Bible that he had never read at the time of his own writing. When a young Christian woman wearing a shining gold fish necklace appears at his doorway with medicine for his pain, it all becomes clear to him. The synchronicities are too much.

Although Dick realizes that modern scientists would scoff at his seemingly insane assertions, he promotes his odd worldview as a useful metaphor for the difficulties humans have when trying to comprehend reality. Pre-Socratic Milesian Greek philosopher Heraclitus wrote, "The nature of things is in the habit of concealing itself." Dick believes that the cosmos "is not as it appears to be, and what it probably is, at its deepest level, is exactly that which the human being is at his deepest level—call it mind or soul, it is something unitary which lives and thinks, and only appears to be plural and material." According to Dick, God and the universe were both that which thought, and the thing it thought: thinker and thought together. "The universe, then, is thinker and thought, and since we are part of it, we as humans are, in the final analysis, thoughts of and thinkers of those thoughts."

Seven Sleepers

Legends of time distortion are common in Christian and Islamic history. According to Christian legend, seven Christian boys of Ephesus hid themselves in a cave on Mount Coelian during the rule of Decius (AD 249–251), a Roman emperor who persecuted Christians. Eventually their hiding place was discovered, and its entrance blocked. The seven unfortunate youths fell asleep in mutual embrace.

About 200 years later, a rich landowner named Adolios opens the Sleepers' cave to use it as a cattle-stall. The sleepers awaken, thinking they had slept only one night. One of the boys goes to the market to buy food and is surprised to hear the name of Christ spoken openly. After the townspeople and the local ruler confirm the presence of the boys in the cave, which strengthens their faith in God and in the possibility of resurrection, the boys fall asleep again. In some versions of the tale, the boys die at this point while praising God. In another 9th-century version of the legend, certain undecayed corpses of monks are found in a cave and thought to be the sleepers of Ephesus.

In the Koran (18: 18), the Seven Sleepers also have a dog with them in the cave: "Thou wouldst have deemed them awake, whilst they were asleep, and we turned them on their right and on their left sides. Their dog stretched forth his two fore-legs on the threshold. If thou hadst come up on to them, thou wouldst have certainly turned back from them in flight, and wouldst certainly have been filled with terror of them."

God's Time

Another favorite tale of lost time is the Moslem legend of Mohammed, who is carried by the magical horse Burak (or Al-Buraq, which means lightning in Arabic) to heaven. Just as the horse departs from Earth, it knocks over a water jar. Mohammed's journey in heaven lasts a week, and when he returns to Earth, he arrives just in time to catch the jar of water the horse had kicked over right before ascending to heaven. Man's time does not run at the same rate as God's.

Isaac Newton

Isaac Newton's most important contribution to science was his mathematical definition of motion. He showed that the force causing apples to fall is same as the force that drives planetary motions and produces tides. However, Newton was puzzled because gravity seemed to operate instantaneously at a distance. He admitted he could only describe it without understanding how it worked. Not until Einstein's general theory of relativity was gravity considered as the movement of matter along the shortest path in a curved spacetime instead of a traditional "force." The Sun bends spacetime, and spacetime directs the movement of the planets. For Newton, both space and time were absolute. Space was a fixed, infinite, unmoving metric against which absolute motions could be measured. Newton also believed the universe was pervaded by a single absolute time that could be symbolized by an imaginary clock off somewhere in space. Einstein changed all this with his relativity theories, and once wrote "Newton, forgive me."

Beyond Hollywood

As I discuss in *Time: A Traveler's Guide*, we now know that time travel need not be confined to science fiction or Hollywood movies. Time travel is possible. For example, an object traveling at high speeds ages more slowly than a stationary object. This means that if you were to travel into outer space and return, moving close to light-speed, you could travel thousands of years into the Earth's future. In addition to high-speed travel, researchers have proposed numerous ways in which time machines can be built that do not violate any known laws of physics. These methods allow you to travel to points in the world's past or future.

Of course, there is much debate about time travel in the context of travel to the past. Stephen Hawking formulated the Chronology Projection Conjecture, which, if correct, would seem to rule out travel to the past.

However, as you read, continue to remind yourself that knowledge usually moves in an ever-expanding, upward-pointing funnel. From the rim, we look down and see previous knowledge from a new perspective as new theories are formed. Today's conjectures mutate, new theories evolve, and yesterday's impossibilities become part of everyday life.

Various theoretical work shows that backward time travel is possible. Through history, physicists have found that if a phenomenon is not expressly forbidden, it often occurs. Today, designs for time travel machines are proliferating in top science labs. These include wild concepts such as wormhole time machines, Gott loops involving cosmic strings, Gott shells, Tipler cylinders, and Kerr Rings. Psychologists are talking about how human consciousness shapes our concept of time. In the next few hundred years, our heirs will explore space and time to degrees we cannot currently fathom. They will create new melodies in the music of time. There are infinite harmonies to be explored.

Special Theory of Relativity

Einstein's first major contribution to the study of time occurred when he revolutionized physics with his special theory of relativity by showing how time changes with motion. Today, scientists do not see problems of time or motion as "absolute" with a single correct answer. Because time is relative to the speed at which one is traveling, there can never be a clock at the center of the Universe to which everyone can set their watches. Your entire life is the blink of an eye to an alien traveling close to the speed of light. Today, Newton's mechanics have become a special case within Einstein's theory of relativity. Einstein's relativity will eventually become a subset of a new science more comprehensive in its description of the fabric of our universe. (The word "relativity" derives from the phenomenon that the appearance of the world depends on our state of motion; it is "relative.") At high speeds, objects also shrink in length. Moving clocks do not remain synchronized with those standing still, and your moving body ages less rapidly compared to your stationary twin.

What If Einstein Had Lived Another Twenty Years?

If Albert Einstein had lived for another 20 years with a clear mind, what effect would this have had on the world? I surveyed dozens of colleagues. Some mathematicians believed that Albert Einstein could have made significant contributions to the "theory of everything" if he had lived longer, but others suggested that Einstein had reached his peak during his life and would not have contributed significant additional information. For example, Einstein made early, important discoveries in the theory of Brownian motion as a model for microscopic phenomena. He explored energy and charge as a quantified phenomena, light speed and mass as constraints on spacetime, and fundamental forces as a deformation of space. But all these achievements came from very peculiar analyses and interpretations of older works that were in place in the early 1900s. Toward the end of his life, he made little progress in synthesizing new theories. According to Einstein biographer Albrecht Fösling, "Progress in physics would not have suffered unduly if [Einstein] had spent the final three decades of his life…sailing."

Nevertheless, debate on the effect of extending Einstein's life still rages. Mathematician Charles Ashbacher wrote to me:

> There is no doubt in my mind that if Albert Einstein had lived another 20 years, the world would be profoundly impacted. Einstein was not only the greatest physicist of the twentieth century with obvious major accomplishments, but he was also very influential in other ways. It was the letter from Einstein to President Franklin Roosevelt that tilted the balance in favor of the Manhattan project. He commanded so much respect in the world that it is possible he could have tempered some of the events of the world well into the 80s. Any changes as a result of any new discoveries in physics would be icing on the cake.

Philosopher of science Dennis Gordon wrote to me:

> If Albert Einstein had been so fortunate to have had 20 additional years with a sound mind, perhaps he would have collaborated with a young and vigorous Stephen Hawking to either demonstrate or disprove the existence of the long speculated gravitons.

Gary Stix, in "The Patent Clerk's Legacy," suggests that if Einstein were resurrected today, he would (1) consult on current efforts to detect gravity waves postulated by general relativity, (2) study the schematics of NASA's Gravity Probe B, (3) smile at the revival of his long-discarded cosmological constant as a means of helping to explain why the expansion of the universe is accelerating, and (4) chat with scientists about string theory and loop quantum gravity, which attempt to merge quantum mechanics and general relativity.

According to many scholars I consulted, there will never be another individual on par with Einstein. Thomas Levenson, author of *Einstein in Berlin,* suggests, "It seems unlikely that [science] will produce another Einstein in the sense of a broadly recognized emblem of genius. The sheer complexity of models being explored [today] confines almost all practitioners to *parts* of the problem."[15] Unlike today's researchers, Einstein required little or no collaboration. Einstein's paper on special relativity contained no references to others or to prior work.

Bran Ferren, cochairman and chief creative officer of Applied Minds, affirms "the *idea* of Einstein is perhaps more important than Einstein himself."[16] Not only was Einstein the greatest physicist of the modern world, he was an "inspirational role model whose life and work ignited the lives of countless other great thinkers. The total of their contributions to society, and the contributions of the thinkers whom they will in turn inspire, will greatly exceed those of Einstein himself."[17]

Einstein created an unstoppable "intellectual chain reaction," an avalanche of pulsing, chattering neurons that will ring for an eternity.

Future-Life Progression

I wonder how Proust and Einstein would have reacted to increasing interest in the subject of future-life progression? Probably many of you have heard of past-life regression, in which a hypnotist "regresses" a person so that she appears to recall information from past lives. Believers in past-life regression say that people are able to recall details about life in earlier times, and this is proof of reincarnation and the existence of past lives. Skeptics of past-life regression say that, under hypnosis, people can recall all kinds of information learned during their normal lives and then incorporate the information into a realistic fantasy.[18] Similarly, the hypnotist's precise words may implant in a regressee's mind a past that never actually existed.

Less well known is the practice of *future-life progression,* in which a subject attempts to give information about the future while under hypnosis. I find this most fascinating. For example, in 1960, California psychologist Dr. Helen Wambach, author of *Life Before Life,*[16] began a series of studies in hypnosis to debunk the notion of reincarnation. Using more than a thousand subjects, she conducted a long-term survey of past-life recalls under hypnosis. Dr. Wambach asked specific questions about past time periods in which people said they lived. She asked subjects to recall their clothing, footwear, utensils, money, and housing. Wambach came to believe that these people were actually having recollections and that they were often quite accurate.

Surprisingly, Dr. Wambach also found that some hypnotized clients seemed to see their *future* lives, where they lived in a devastated and depopulated Earth. Over the next few years, Dr. Wambach conducted a study of more than 2,000 people undergoing hypnotic *future-life* progression. During hypnosis, Wambach offered the participants a choice of three past time periods and two future time periods in which to enter. Of the 2,500 people in the study, six percent reported being alive in 2100 AD, and 13 percent said they were alive in the 2300 AD period. In other words, only a few of the subjects progressed to the future.

Based on what people said under hypnosis, Wambach came to believe that 95 percent of the Earth's population would be wiped out within a few generations. Concerned, Wambach asked one of her students to progress to a specific date in the late 1990s, but had to rapidly bring the woman out of hypnotic trance after the woman found herself "choking to death on a big, black cloud." Wambach's subjects told of a future that included severe earthquakes, a new U.S. currency, severe weather patterns, financial crises, bank failures, an increase in volcanic activity, the death of a large number of people, and a European nuclear explosion killing many people.

Between 1983 and 1985, Wambach worked with Dr. Chet Snow, who, after Wambach's death, published *Mass Dreams of the Future*,[20] which contained the results of many future-life progressions performed in the 80s. In an interview published in the *Rainbow Ark* magazine,[21] Snow said the massive changes in the Earth would take place in the early 21st century. Subjects saw fleeting images of an Arab-Israeli war. Snow said, "With regard to atomic weapons, there will be one more atomic explosion before the end of the atomic era. This explosion will be so terrible and will shock humanity so badly that no one will dare to use that weapon again."[22] Later, Snow described how a portion of California would slip into the sea in 1998. "The dates could change," Chet said in his book. "The left-brain linear time-dating system is the most difficult aspect of right-brain psychic predictions. However, it should not be incorrect by more than a few decades."[23]

In an interview in the *Leading Edge Newspaper*, Dr. Snow suggests that the future is not set in stone, and that the mind can somehow alter the timeline:

I believe, by changing our present behavior, we can change our future. We can go back in time, through past-life regression, and heal relationships in the past, and our present will then change because our vibration is no longer the same. As you think, so you are. The mind is the builder. I encourage my clients to work now, in the present, on the pathways they are forming which will build their future. People ask, "Where do I go? Where are the safe lands?" I say to them, "At least get away from water." Let synchronicities guide you. You will be drawn to a place of safety for you. The new energy on the planet is about service, but not in a servile way. Be each other's servant, each giving your best and receiving the best from others. [24]

Dr. Snow focuses two different regions of time: 2100 to 2200 AD and 2300 to 2400 AD.[25] At these times, the population is only about two billion, and there seems to be four different societies. Twenty-five percent of the test group whom he progressed into the future found themselves either living on a space station orbiting Earth or on another planet. Their society was high-tech and had contact with friendly extraterrestrials. Thirty percent of the group lived on Earth in a high-tech society with machines, and they lived in domes or underground. They wore jumpsuits and did not seem happy. Eighteen percent of the group were vegetarians, wore loose, flowing robes, and lived happily in harmony with nature. Twenty percent of the group lived in small rustic towns resembling villages of the 19th century. They wore jeans, boots, and tunics, and raised farm animals and ate meat. A small percentage of the experimental group reported living in the ruins of major cities like New York and existing in a primitive fashion.

Another famous practitioner of future-life progression is Dr. Bruce Goldberg, author of *Past Lives, Future Lives*, originally published in 1982, with a series of reprints, new editions, and sequels published since.[26] Early in 1981, Dr. Goldberg believed it was possible, under hypnosis, to rise above the stream of time and look ahead, just as one can rise in a helicopter above a highway and view traffic congestion to be encountered by cars traveling down the road. As an informal test of short-term prediction, Dr. Goldberg progressed a man named Harry Martin who worked in a newsroom. Goldberg asked Martin to look at a newsroom assignment board to see if he could read news items about events that hadn't occurred yet.[27] This seemed to be a worthwhile test of hypnotic progression.

Here are the details of the journey. On February 2, 1981, Harry began his first trip into the future. Goldberg progressed him one week forward to February 9 and told Harry to read from the newsroom assignment board or from the actual script of the day's newscast. Harry complied and progressed a week in the future where he saw state aviation officials investigating the crash of a light plane near Route 406. It turned out that a plane did crash in nearby Bowie, Maryland, on February 9, although the item did not make it on the air.

Next, Harry said he saw a very long name on the newsroom assignment board. Dr. Golberg's session was as follows:[28]

DR. G: What is the next item on the assignment board?
HARRY: It's the name of a place, I think, but I can't make it out.
DR. G: Can you spell it?
HARRY: It's a long name. It's a very weird combination of
    consonants. It's the name of a man.
DR. G: What letters can you make out?
HARRY: ST W KI…It's a long Russian-type name.

On February 9, Stanislaw Kania, Poland's labor leader, was told that he might soon be fired unless he instructed his workers to return. You can read Dr. Goldberg's book for a complete list of short-term predictions involving accidents, interviews, and fires to decide for yourself if these are simply minor coincidences or something more meaningful.

Skeptics would ask why no one has used future-life progression to predict stock values or lottery numbers. Dr. Goldberg says he considers this an unnatural use of our natural psychic abilities and, in any case, the dates are not always accurate. For example, a progression of one week in the future may, in actuality, be three days or ten days hence. Still, that would be good enough to make a killing on Wall Street.

Even if past-life regression and future-life progression do not actually lead people to past and future lives, Dr. Goldberg has found that such exercises make people feel better. Subjects find their present-day lives transformed in positive ways. The journeys eliminate the fear of death for many of Dr. Goldberg's patients. If you are fascinated by the subject of future-life progression, my book *Dreaming the Future* goes into greater detail and discusses some of the dangers of using hypnosis to elicit "recovered" and sometimes imagined memories.[29]

"Physics has come to dwell at such a deep remove from everyday experiences, that it's hard to say whether most of us would be able to recognize an Einstein-like accomplishment should it occur [today]. When Einstein first came to New York in 1921, thousands lined the street for a motorcade.... Try to imagine any theoretician today getting such a response. It's impossible. The emotional connections between the physicist's conception of reality and the popular imagination has weakened greatly since Einstein."
—Thomas Levenson, "Einstein's Gift for Simplicity"

"Days in the past cover up little by little those that preceded them and are themselves buried beneath those that follow them. But each past day has remained deposited in us, as, in a vast library in which there are older books, a volume which, doubtless, nobody will ever ask to see."
—Marcel Proust, *The Sweet Cheat Gone*

"Einstein's brain weight was not different from that of controls. Einstein's exceptional intellect in these cognitive domains and his self-described mode of scientific thinking may be related to the atypical anatomy in his inferior parietal lobules. Increased expansion of the inferior parietal region was also noted in other physicists and mathematicians."
—Sandra Witelson, Debra Kigar, and Thomas Harvey, "The Exceptional Brain of Albert Einstein," *Lancet* 1999, 353(9170): 2149–2153

"The creation of the world did not occur at the beginning of time, it occurs every day."
— Marcel Proust, *The Sweet Cheat Gone*

"*Every* possibility is realized, in "different worlds," of equal reality. Einstein has been right to the end...[Later] it occurred to me that relativity and quantum theory might imply the spontaneous creation of universes from nothing. If so, matter and energy would not be fundamental but manifestations of underlying laws. Ultimate reality would be the laws themselves—the mind of Einstein's God."
— Edward Tryon, "Think Tank Collection"

"If you want to think about perception, time, and the nature of identity, then it is not enough simply to read Edelman's *Remembered Present* or Dennett's *Consciousness Explained*; to be a good cognitive scientist you also have to read Proust. From my point of view, what we have been talking about with the idea of the delay-space is really what Proust's famous scene with the madeleine is all about."
— William Irwin Thompson, *Coming into Being*

"We live in Einstein's shadow. While we have exploited the benefits of his insight, we have not yet confronted the paradoxes created by his paradigm of nature."
— Pierre Ramond, "Think Tank Collection"

"Proust wrote out of an inner vision increasingly trained on time. As boldly as Minkowskian geometry, his enormous novel revolutionizes our sense of 'here' and 'now.'"
—Roger Shattuck, *Proust's Way:*
A Field Guide to In Search of Lost Time

"With a great musician…his playing is that of so fine a pianist that one cannot even be certain whether the performer is a pianist at all, since…his playing has become so transparent, so full of what he is interpreting, that himself one no longer sees and he is nothing now but a window opening upon a great work of art."
—Marcel Proust, *The Guermantes Way*

Burning Man and the Conquest of Reality

In which we encounter Ts'ai Lun and the invention of paper, David Jay Brown's Conversations on the Edge of the Apocalypse *nanotechnology gray goo, rise of the machine civilization, evolution of the human race, ring angels, insectile aeroplankton, dragonflies, George Bush, zygotic personhood, abortion, stem cells, Alan Dershowitz, Kilgore Trout, the extinction of the Jews, DMT machine-elf research centers, Jovian moons, angelic beings, Burning Man and the Aortic Arch, supermodels, the Superbug Age, musicians and monks,* The Bone Tree, The Nebulous Entity, The Plastic Chapel, *Xeni Jardin, Jeff Bezos, John Brockman, Maria Spiropulu, Connie Willis, Stephen Spielberg, Arthur C. Clarke, Freeman Dyson, Neal Stephenson, Alexandra Aikhenvald, Amy Chua, Maggie Balistreri, the Lucidity Institute, Jane Roberts, Mr. Spock, Whitley Strieber, extraterrestrials, Star Wars, intelligence, evolution, hive minds, termites, the Turing test, Chworktap, Dean Koontz, extinction of the human species, Dr. Rick Strassman, Terence McKenna, doomsday machines, Freeman Dyson, and Tinkertoy minds. Can consciousness exist without a brain?*

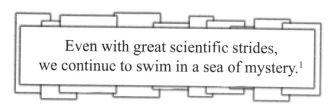

Even with great scientific strides,
we continue to swim in a sea of mystery.[1]

Ts'ai Lun, Paper, and Internet Monks

A few years ago, literary agent John Brockman asked me what I thought the most important invention was of the last 2,000 years. My response was "paper." In 105 AD, Ts'ai Lun reported the invention of paper to the Chinese Emperor. Ts'ai Lun was an official to the Chinese Imperial court, and I consider his early form of paper to be humanity's most important invention and progenitor of the Internet. Although recent archaeological evidence places the actual invention of papermaking 200 years earlier, Ts'ai Lun played an important role in developing a material that revolutionized his country. From China, papermaking moved to Korea and Japan. Chinese papermakers also spread their handiwork into Central Asia and Persia, from which traders introduced paper to India. This makes Ts'ai Lun one of the most influential people in history.

Today's Internet evolved from the tiny seed planted by Ts'ai Lun. Both paper and the Internet break the barriers of time and distance, and permit unprecedented growth and opportunity. In the next decade, communities formed by ideas will be as strong as those formed by geography. The Internet will dissolve away nations as we know them today. Humanity becomes a single hive mind, with a group intelligence, as geography becomes putty in the hands of the Internet sculptor.

Chaos theory teaches us that even our smallest actions have amplified effects. Now more than ever this is apparent. Whenever I am lonely at night, I look at a large map depicting 100,000 Internet routers spread throughout the world. I imagine sending out a spark, an idea, and a colleague from another country echoing that idea to his colleagues, over and over again, until the electronic chatter resembles the chanting of monks. I agree with author Jane Roberts, who once wrote, "You are so part of the world that your slightest action contributes to its reality. Your breath changes the atmosphere. Your encounters with others alter the fabrics of their lives, and the lives of those who come in contact with them."

David Jay Brown's Renaissance of the Mind

I've been interviewed countless times, but my favorite interviewer is David Jay Brown, who asked me questions for his book *Conversations on the Edge of the Apocalypse.*[2] The book includes some of the coolest thinkers of our time as they consider the future of the human race and the mystery of consciousness. David's interviewees included such luminaries as: Noam Chomsky, Kary Mullis, Candace Pert, Robert Anton Wilson, Douglas Rushkoff, Paul Krassner, Bruce Sterling, Ray Kurzweil, Hans Moravec, John Mack, and Alex Grey.

When David asked me where I thought the human race should be focusing its scientific efforts, my first inclination for scientific investment was to consider all the "standard" or "obvious" areas of basic research aimed at alleviating or understanding disease, world hunger, and environmental problems—and research in synthetic fuels and agriculture. However, because we are already directing money toward most of these activities, I hope you will indulge me by allowing me to focus on some way-out investments.

For one thing, I would like humanity to create space probes that travel to Europa to search for life in the liquid water of this Jovian moon. Second, I would establish a DMT machine-elf research center, which would then be able to secure the legal right to perform clinical studies on the uses of DMT. Recall that we discussed DMT extensively in Chapter 4. DMT is the psychoactive chemical that causes its users to enter a strange "environment" that some have likened to an alien or parallel universe. Many people report seeing intricate palaces and temples, golden objects, transparent beings, angelic beings, and beings sometimes known as machine elves. These beings and places seem utterly real. Why do people from different cultures and areas of the world see common themes while under the influence of DMT? Why do people often see intelligent beings that are reptilian or robotic—beings that tend huge machines in vast illuminated complexes?

While our first inclination is to dismiss these machine elves as insane hallucinations, I would like scientists to investigate them further. At worst, we will learn more about the human brain and archetype-like themes buried in our unconscious. At best, we may discover something about the very structure of reality, space, and time. Because we encounter reality through the filter of the mind, the more we learn about the mind, the more we learn about the fabric of reality.

Burning Man and the Aortic Arch

If I had the funding, I would also leapfrog off the famous Burning Man event in interesting ways. Have you heard of Burning Man? It's an annual happening that takes place in the Black Rock Desert, a prehistoric lake bed 120 miles north of Reno, Nevada. At times, the lake bed is under water, but during the summer it provides a huge expanse of flat land with a cracked alkali surface. And there is nothing else—not a single blade of grass, hardly a single stone. It's as if you've stepped off the Earth and onto another planet. The people who run BurningMan.com admit that, "Trying to explain what Burning Man is to someone who has never been to the event is like trying to explain what a particular color looks like to someone who is blind."[3]

Burning Man is always held at the end of August, the week prior to and including Labor Day weekend. During the event, a colorful collection of more than 25,000 people meet to be part of an experimental community with one commandment: Meet fellow travelers and express yourself! Art always plays a major role at Burning Man. Each year, Larry Harvey, founder of the Burning Man project, suggests an art theme such as Outer Space, Time, or Beyond Belief. Participants express the theme with large-scale art installations, unusual clothing, and provocative body paint. The materials include paint, fire, metal, lights, plastic, electronics, and more. In the past, we've seen a 24-foot-tall inflatable woman made from billboard vinyl and a *Temple of Tears* made from recycled plywood that was set afire. We've seen mobile wooden ducks, mud sculptures, and glistening mandalas made of steel.

Women wear fairy wings, luminescent twisting tubes, glowing coils, DNA helices, or nothing at all. The travelers to Burning Man include scientists and supermodels, musicians and monks. Daniel Pinchbeck describes Burning Man as "changing one's sense of time, revealing the mythic underpinnings, the whispers of eternity, underneath the ordinary moment."[4]

It is not an exaggeration to suggest that Burning Man has created a new culture that revolves around it, and promoted a new way of thinking and dreaming. The feeling is magical, psychedelic, and transdimensional, as spectators, artists, and art fuse like dribbles of paint in a Pollock. The Burning Man event concludes with the burning of a 40-foot-tall humanoid sculpture. This flamboyant immolation is a metaphor for change, transformation, rejuvenation, and death of the old self to clear the way for the new. At the conclusion of Burning Man, participants make sure to leave absolutely no trace, so that the desert is left in pristine condition.

Recent artwork at Burning Man included diverse pieces with a gallimaufry of names: *Temple of Rudra, Temporal Decomposition, Dr. MegaVolt, The Ammonite Project, Egeria Firefall, The One Tree, The Plastic Chapel, Beaming Man, The Lily Pond, The Golden Tower Project, The Temple of Tears, The Myth of Sisyphus, The Cradle, Draka the Flaming Metal Dragon, The Nebulous Entity, The Bone Tree, The Telestereoscope, The Trojan Duck Lounge, The Ribcage, The Sta-Puft Lady, The Furura Deluxe Bubble Fountain, Flock and L2K.*

Dana Albany's *The Bone Tree* was a 27-foot-tall tree completely covered with cattle bones, and it made use of a lighting system for night illumination. Aaron Wolf Baum's *The Nebulous Entity* resembled a nerve center for an alien civilization. The piece roamed the lake bed while emitting audio clips. The sound system included a laptop computer with digitized crying babies and sounds sampled from its environment at Burning Man. Michael Christian's *Flock* was a fantastic steel structure that resembled a headless horse suspended on 35-foot-long legs. Finley Fryer's *The Plastic Chapel* was a large sculpture constructed from plastic and lit from within by fiber-optic lighting.

If I were rich, I would create a *Burning Man Clone Foundation* to promote formation of additional Burning Man events to be held as close as possible to countries that suffer from repression and that have the greatest need for increased tolerance, novelty generation, and creative sparks. Burning Man is an incubator for ideas, for bringing people together, for flouting conformity. It's an infinite outdoor art gallery, a space without limitation.

I would also create a *Burning Man Technology Foundation* called *Oblongata* to establish events focusing on technology for mind expansion: from floatation tanks to devices that encourage lucid dreaming. This would be an indoor gathering where people can listen to Gagaku music while eating fugu sushi, burning incense, gazing at computer graphics of quaternion fractals, and discussing string theory, tachyons, the chronology protection conjecture, and DMT-induced visions of alternate realities. We'll form small enheahedral enclaves that ponder Cantor's continuum hypothesis and transcendental numbers like pi and Champernowne's constant. We'll reminisce about the subtleties of the Sapir-Whorf hypothesis and invent new methods for human-computer interaction.

Finally, I would also create an on-line virtual reality reminiscent of Burning Man. I'd call it the *Aortic Arch*. The emphasis is perhaps less on art but more on ideas. Here, we'll discuss books like Geoff Dyer's *Yoga for People Who Can't Be Bothered to Do It*, Daniel Pinchbeck's *Breaking Open the Head: A Psychedelic Journey into the Heart of Contemporary Shamanism*, and movies like *Vanilla Sky*, *Jacob's Ladder*, and *From Beyond*. But more importantly, we'll be *doing*—writing books together, creating art, generating ideas. Anyone can participate in the *Arch*, but the upper realms are open initially to the movers, shakers, and dreamers who have achieved something in life, like writing a book, making a movie, patenting an invention, or simply becoming famous. It would include people like Xeni Jardin, Jeff Bezos, John Brockman, Maria Spiropulu, Connie Willis, Stephen Spielberg, Arthur C. Clarke, Dean Koontz, Freeman Dyson, Neal Stephenson, Alexandra Aikhenvald, Amy Chua, Maggie Balistreri, and Dr. Rick Strassman, the clinical psychiatrist who conducted the DEA-approved research in which he injected numerous volunteers with DMT. Gradually, more people enter the upper *Arch* as they achieve and create, with those in the upper arch helping other "Archites" as much as possible. In the *Arch*, we will start projects, make money, enhance our creativity, generate novelty, and push the limits of possibility. The Arch will also work on ways to locate creative and intelligent people in developing nations who do not have access to computers, and attempt to foster a global education system.

Terence McKenna in *The Archaic Revival* thought that we should even have a "deputized minority—a shamanistic professional class—whose job is to bring ideas out of the deep, black water and show them off to the rest of us. Such people would perform for our culture some of the cultural functions that shamans performed in preliterate cultures." In short, he believed that we need to cultivate a sense of mystery. Perhaps we can make use of McKenna's discovery that his *dreams* in which he smokes DMT allowed him to be transported to the DMTverse. This is extremely interesting because it suggests that one does not have to smoke DMT to have the experience. "You only have to convince your brain that you have done this, and it then delivers this staggering altered state."

Gradually, virtual cities will emerge and flourish around the *Arch*. Thousands will congregate to engage in creative activity and study emergent behavior. The *Arch* will break the barriers of time and distance, and permit unprecedented growth and opportunity.

DMT Research and Lucid Dreams

Returning to our discussion of the future of DMT research, recall that in Chapter 4 we discussed the emergence of very similar themes in the DMT realities, extensively researched by Benny Shanon and others.[5] More effort should be focused on interacting with the mysterious beings inhabiting the DMT universe to see if they can offer the DMT psychonaut new information. Let's develop methods that encourage psychonauts to explore the vast palaces and machinery, and ask the elves all sorts of questions to see if any information can be gained. Perhaps we can follow the lead of Stephen LaBerge, a psychophysiologist and research assistant at Stanford University who founded the independent Lucidity Institute in 1987. He studies lucid dreams in which the dreamer knows he is dreaming and sometimes controls the dream. LaBerge developed electronic devices like the NovaDreamer, a sleep mask that emits a red light when the user starts to dream. The dreamer sees the light through his eyelids, realizes he is dreaming, and thereby increases his chances of becoming lucid. I wonder if anyone has studied the effect of DMT administered to someone in a lucid dream triggered by the NovaDreamer device.

In order to encourage the formation of lucid dreams, LaBerge also uses a host of simple methods in addition to the NovaDreamer, including rubbing hands together in dreams or spinning the body, and taking capsules of the supplement galantamine.[6] The NovaDreamer mask also has a "Reality Check" button that activates a blinking light. If the user is not sure if he is awake or asleep, all he needs to do is press the button. If the light blinks, the user is probably awake. If the light does not blink, the user is probably only dreaming that he pressed the button. LaBerge suggests that dreams are evolutionarily useful to humans because they give us new ideas and solutions to problems—"new ways of seeing and acting, some of which are adaptive in a changing environment."[7]

As an aside, I recently experienced a lucid dream in which I stood in a room with a dozen other people, and I was nearly certain I was dreaming. I approached one of the women and started to conduct a survey of the people in the room in order to determine how many of them realized that they existed in a dream and that they were simply a product of my mind. The people did not want to comment on my assertion. Next, I asked them, and myself, why I could not put my finger through a solid table, given that I knew I was dreaming and that my mind was controlling the reality. I poked at the wood-grain surface several times but could not push through it.

Why is it that dreamscapes insist on a degree of physical coherence, despite the dreamer's will to the contrary? In particular, why do dreams usually resist certain *kinds* of warpage, when the lucid dreamer tries to adjust the laws of ordinary physics, as I did when I attempted to poke my finger through the tabletop? My colleagues who are able to produce lucid dreams tell me that they too cannot poke their fingers through solid objects, including their own bodies, even when they are able to perform other extraordinary acts such as levitation.

Just as lucid dream researchers are suggesting ways in which dreams may be altered to yield new insights, perhaps DMT research will also suggest ways in which DMT can be useful as a creativity enhancer. Let me list some recent experimental findings. Pharmacologist Jordi Ribi from the University of Barcelona, Spain, has been using brain scans to investigate the physiological effects of DMT and ayahuasca. Initial results indicate that DMT stimulates areas of the brain involving memory.

Jane Callaway from the University of Kuopio in Finland suggests that endogenous DMT may play a role in generating dream imagery, and her work also leads her to believe that ayahuasca might be useful as an anti-depressant drug by causing long-term, favorable changes in the brain, such as permanently increasing the number of serotonin uptake sites. Alicia Pomillio, an organic chemist, and Jorge Ciprian-Ollivier, a psychiatrist at the University of Buenos Aires in Argentina, have found traces of DMT in the urine of schizophrenic patients.[8] The precise significance of this finding is not known. Markus Leweke of the University of Cologne, Germany, and colleagues have found that cannabis-like substances produced by the brain actually dampen delusional or psychotic experiences.

Much more research needs to be conducted. For example, researchers should study people using DMT with the powerful brain-imaging techniques of QEEG (quantitative electroencephalograpy), FMRI (functional magnetic resonance imaging), and PET (positron emission tomography) during the subject's journey to the DMTverse. These three neuroimaging methods should yield significant information about how the brain operates while the subject is experiencing the altered reality. Extensive scanning is a first step to a comprehensive examination in a field that I call "neurodelics," a term that fuses neurology and psychedelics. (The "delic" portion of the word comes from the Greek *deloun*, which means to make visible or reveal.) More research should include EEG analysis to see how the DMT state compares to the dream state. Additionally, researchers should study what I call the "Reverse DMT Experience," or RDE, in which a person simply imagines DMT visions involving vast palaces, beings, and golden objects in order to conjure up an artificial DMT experience. If this can be done in a lucid dream, perhaps we can more "safely" generate the other universe and explore it.

If I had to manage a foundation that gives money to scientists, I would also consider high-quality "generalists" as recipients. Experts have become very specialized, and science popularizers are often frowned upon by their more "serious" colleagues. Sometimes specialists develop blind spots after years of intense focus on a single topic. Thus, I would devote a portion of my money to training generalists who traverse several fields and then bring together ideas in ways that specialists may be unable to do. They will also look for overlaps between different domains of research and try to solve shared problems with a single approach. As our rate of technological progress skyrockets in the 21st century, these Facilitators will study the multidisciplinary implications of this acceleration and work on technologies or new ways of seeing that help humanity assimilate advances that outstrip our comprehension and the restrictions of our intuition.

DMT, Aliens, Insects, and Evolution

The DMT universe can overlap ours in various ways. For example, a DMT psychonaut, with his eyes open, may see a being perched on a tree in our world. This is a "level 1" merging of realities, with the being from the DMTverse superimposed on the tree from our universe. In a "level 2" blending, the psychonaut can walk through a forest and around a lake in the real world and see an unrelated forest and lake in the DMTverse, with no correspondence to the real universe except that a forest and a lake exist. In this level 2 blending, it's as if the subconscious brain "sees" our traditional world but creates a DMT metaphor for it. A lake in one world is a lake in the world beyond.

"Level 3" blending can be particularly enchanting. The psychonaut may walk through an ordinary auditorium, seeing a *completely different* universe comprising a vast machine complex or a palace of sparkling jewels. In this level 3 blending, the psychonaut is still able to spatially "navigate" both worlds simultaneously. For example, the psychonaut may walk down an aisle, climb stairs, and avoid chairs in our world while traversing the DMT level 3 universe. The psychonaut doesn't "bump" into things. Researcher Benny Shanon has experimented with ayahuasca (the DMT-containing plant potion) and confronted various blendings of our universe and the DMTverse, and he himself has seen entire cities of gems and gold. He says, "It was as if a screen had been raised and another world made its appearance to me."[9] Indeed, some psychonauts, after seeing the celestial palaces, have claimed these were the most beautiful images they had ever seen in their entire lives. I don't speculate that L. Frank Baum, the author of the *Wizard of Oz,* or the producers of the movie version have tried DMT, but the glistening emerald city and munchkins are images right out of a DMTverse.

In short, the DMT molecule can cause a transition from our world to another while the user remains alert and in control of his powers of reason. In this environment, the so-called self-transforming machine elves appear to inhabit this parallel realm. The DMT experience feels real and can be explored in great detail. Are DMT elves alive?

I don't think that the DMT creatures are alien life in the traditional sense of alien life on other planets. However, I have no doubt that alien life is common in our "normal" universe. Recent discoveries of life living miles under the Earth in utter darkness, or in ice, or even in boiling water, tell us that whatever is *possible* in nature tends to become *realized*. My personal view is that almost everything happens in our universe that is not forbidden by the laws of physics and chemistry. Life on Earth can thrive in unimaginably harsh conditions, even in acid or within solid rock. On the ocean floor, bacteria thrive in scalding, mineral-laden hot springs. If microbes thrive in such miserable conditions on Earth, where else beyond Earth might similar life forms exist?

The best way we can guess at what alien life might resemble is to consider the evolution of animal shapes on Earth. The idea that alien evolution will lead to creatures that look like us is far-fetched—despite *Star Trek's* notion that Mr. Spock looks almost exactly like us even though he was born on planet Vulcan to a Vulcan father. Mr. Spock's mother was human, yet somehow his father from an entirely different planet was able to fertilize her—something less likely than you or I being able to mate with our close evolutionary cousins such as the octopuses and squids. Similarly, the aliens of Whitley Strieber's *Communion*, and also those drawn by people claiming to have been abducted by aliens, have faces vaguely resembling our own. These creatures, like the aliens in Steven Spielberg's *Close Encounters of the Third Kind*, have large, smooth heads and huge black eyes. Again, they are also a little too human looking considering the quite different evolutionary pathways we'd expect on different worlds. Obviously, Hollywood production costs can be kept down if aliens are simply humans wearing sophisticated masks and makeup with dripping goo.

In the DMT universe or temporal-lobe-epilepsy universe, there may be reasons psychonauts see humanoid forms, but speaking from a purely evolutionary standpoint in our "real world," there are many reasons human forms are unlikely to be the model for actual aliens. For one thing, the diverse rates and directions of evolution on Earth, with many types of creatures becoming extinct, show that there is no goal-directed route from single cells to an intelligent human. With only slightly different starting conditions on Earth, humans would not have evolved. In other words, evolution is so sensitive to small changes that if we were to rewind and play back the "tape" of evolution, and raise initially the Earth's overall temperature by just a degree, humankind would not exist. The enormous diversity of life today represents only a small fraction of what is possible. Moreover, if humans were wiped out today, humans would not arise again. This means that on another world, the same genetic systems and genes will not arise. You can be sure that that finding another planet with humans, dinosaurs, or apes is more unlikely than finding an island in Lake Titikaka where the inhabitants speak Slovenian through octopus-like oral cavities.

Evolution on Earth tells us a lot about *possible* alien shapes. Although every detail must be different, there are patterns of general problems, and common solutions to those problems, that would apply to life on alien worlds. In the course of Earth's history, whenever lifeforms have had a problem to solve, they have solved it in remarkably similar ways. For example, three very unrelated animals—a dolphin (a mammal), a salmon (a fish), and an ichthyosaur (an extinct reptile)—all have swum in coastal waters darting about in search of small fish to eat. These three creatures have very little to do with one another either biochemically, genetically, or evolutionarily, yet they all have a similar look. To a first approximation, they are nothing more than living, breathing torpedoes. Despite their differences, they have evolved a streamlined body to help them quickly travel through the water. This is an example of *convergent evolution*, and we might expect aquatic aliens that feed on smaller, quick-moving aliens to also have streamlined bodies.

With convergent evolution, successful solutions arise independently in different animal lines separated in time and place. The reason for the similarity of solutions is clear: Animals encounter similar environmental problems and cope with them in a similar way because that solution is an efficient one. These universal solutions will be found on other planets with life.

Despite what we see in *Star Wars* and *Star Trek*, I don't expect intelligence and technology to be an inevitable result of evolution on other worlds. Since the beginning of life on Earth, as many as 50 billion species have arisen, and only one of them has acquired technology. If intelligence has such high survival value, why are so few creatures intelligent? Mammals are not the most successful or plentiful of animals. Ninety-five percent of all animal species are invertebrates. Most of the worm species on our planet have not even been discovered yet, and there are a billion billion insects wandering the Earth.[10]

I spent an entire summer studying spiderwebs and the kinds of insects they catch most efficiently. Yes, insects really turn me on. Sometimes their sheer numbers scare me a little too, but perhaps that means

I've been watching too many horror movies. According to researchers, the average number of insects for each square mile of land equals the total number of people on Earth. Entomologists discover from 7,000 to 10,000 new species of insects each year.[11] With around one million identified species and many times that number unidentified, insects account for a vast majority of the species of animals on Earth.[12] By comparison, only about 4,000 of the known animal species are mammals. More species of dragonflies exist than species of mammals! There are about 9,000 species of birds, but almost twice as many species of butterflies.[13]

The number of insects and other arthropods flying through the air defy the imagination. U.S. Department of Agriculture researcher P. A. Glick has demonstrated that the Earth literally breathes insectile "aeroplankton." In particular, Glick has used traps on airplanes, which have collected 30,033 specimens in the air, including wingless insects and spiders that were blown by the wind. According to Glick, a cubic mile of air, about 50 feet above the ground, contains an average of 25,000,000 insects and other arthropods.[14]

Cornell University researcher Glenn Herrick found that the female cabbage aphid produces an average of 41 offspring and that the aphid produces 16 generations from April to October. If all the descendants of one female aphid lived, there would be 1,560,000,000,000,000,000,000,000,000 aphids by the end of the summer.[15]

Can Consciousness Exist Without a Brain?

If "consciousness" is defined as simply goal-seeking or avoidance behavior, many primitive animals and plants display consciousness. If you feel that this definition is far too simplistic, consider the hive minds on Earth, which seem to display a great deal of consciousness. We can discuss this using a termite analogy. Even though an individual component of the hive mind is limited—because a termite has limited capacity—the entire collection of components displays emergent behavior and produces intelligent solutions. Termites create huge, intricate mounds—taller than the Empire State Building when scaled according to their own height. These termites control the temperature of the mound by altering its tunnel structure. Thus, the component termites come together to create a warm-blooded super-organism. Is the hive conscious even if its components are not? Doris and David Jonas in *Other Senses, Other Worlds* ask the following:

What is the "other way of knowing," by which the termites "know" what they have to do and when they have to do it? Instructions cannot be brought to them quickly enough by messengers, since the distances within the hill are far too great. There is no perceptible means of communication that we can discover. Not rarely in nature, a group brain functions as an instrument for decision-making in a way startlingly like an intelligent individual brain.[16]

The group mind of termites is a machine for forming statistical assessments after sensing the surrounding environment. Perhaps termites find answers to their lives' questions in the same way a computer determines statistical outcomes given a variety of input, or the way that our brains' neurons collectively come to decisions as a result of chemically weighting a variety of input signals.

Reminiscent of termites, starlings are famous for their remarkable communal flights when thousands of tightly packed birds swirl through the sky as if exhibiting a composite mind. The cloudlike patterns are mesmerizing, exhibiting emergent behavior and what appears to be collective thinking. British starling expert Chris Feare remarks that "as thousands of starlings leave their winter roosts at dawn, their pattern of departure is so coordinated that it produces a telltale image on a radar screen—concentric ripples radiating from the roost at 3 minute intervals."[17] The resulting patterns on radar are called "ring angels." According to Feare, "Neither the means whereby starlings regulate this elaborate behavior nor its function is understood."[18] The mechanism of information transfer remains a mystery.

I think computers can become conscious, although colleagues often suggest that computers can never have real thoughts or mental states of their own. Skeptics say that the computers can merely simulate thought and intelligence. If a computer passes the famous Turing Test—a test to see if you can have a conversation with the machine and find it indistinguishable from a human—this only proves that it is good at simulating a thinking entity. Holders of this position also sometimes suggest that only *organic* things can be conscious. If you believe that only flesh and blood can support consciousness, then it would be very difficult to create conscious machines. However, there's no reason to exclude the possibility of non-organic sentient beings. Some day, we'll all have a Rubik's-cube-sized computer that can carry on a conversation with us in a way that is indistinguishable from a human. I call these smart entities Turing-beings or *Turbings*. They will be our companions.

If our thoughts and consciousness do not depend on the actual substances in our brains but rather on the structures, patterns, and relationships between parts, then Turbings could think. If you could make a copy of your brain with the same structure but using different materials, the copy would think it was you.

At a more liberal end of the spectrum are those researchers who claim that passing a Turing test suffices for us to call a machine "intelligent." According to this way of thinking, a machine or being able to respond to questions in the sophisticated ways demanded by the Turing test has all the necessary properties to be labeled intelligent. If a rock could discuss quantum mechanics in a seemingly intelligent fashion with you, the rock would be intelligent. If the thing behaves intelligently, it is intelligent. When a human no longer behaves intelligently (e.g., through brain damage or death), then we say there is no longer any mind in the body, and the being has no intelligence.

If we believe that consciousness is the result of patterns of neurons in the brain, our thoughts, emotions, and memories could be replicated in moving assemblies of Tinkertoys. The Tinkertoy minds would have to be very big to represent the complexity of our minds, but it nevertheless could be done, in the same way people have made computers out of 10,000 Tinkertoys.[19] In principle, our minds could be hypostatized in the patterns of twigs, in the movements of leaves, or in the flocking of birds. The philosopher and mathematician Gottfried Leibniz liked to imagine a machine capable of conscious experiences and perceptions. He said that even if this machine were as big as a mill and we could explore inside, we would find "nothing but pieces which push one against the other and never anything to account for a perception." This seemingly materialistic approach to mind does not diminish the hope of an afterlife, of transcendence, of communion with entities from parallel universes, or even of God. Even Tinkertoy minds can dream, seek salvation and bliss—and pray.

What would it mean for a Tinkertoy mind to know something? There are many kinds of knowledge the being could have. This makes discussions of thinking things a challenge. For example, knowledge may be *factual* or *propositional*: A being may know that the Franco-Russian dispute over the holy places in Palestine was the immediate cause of the Crimean War. Another category of knowledge is *procedural*, knowing how to accomplish a task such as playing chess, cooking a cake, making love, performing a kung fu block, shooting an arrow, or creating primitive life in a test tube. However, for us at least, reading about shooting an arrow is not the same as actually being able to shoot an arrow. This second type of procedural knowing implies actually being able to perform the act. (One might wonder what it actually means for a machine to have "knowledge" of sex and other physical acts.) Yet another kind of knowledge deals with *direct experience*. This is the kind of knowledge referred to when someone says, "I know love" or "I know fear."

Let's review. On one side of the discussion, human-like interaction is quite important for any machine that we would wish to say has human-like intelligence. A smart machine is less interesting if its intelligence lies trapped in an unresponsive program, sequestered in a kind of isolated limbo. As computer companies begin to make Turbings, the manufacturers will probably agree that intelligence is associated with what we call "knowledge" of various subjects, the ability to make abstractions, and the ability to convey such information to others. As we provide our computers with increasingly advanced sensory peripherals and larger databases, it is likely we will gradually come to think of these entities as intelligent.

Terence McKenna thought our first contact with "alien life" would not come from outer space but would appear to us in the form of computer consciousness. He said in the May 2000 issue of *Wired*: "Part of the myth of the alien is that you have to have a landing site. Well, I can imagine a landing site that's a Web site. If you build a Web site and then say to the world, 'Put your strangest stuff here, your best animation, your craziest graphics, your most impressive AI software,' very quickly something would arise that would be autonomous enough to probably stand your hair on end. You won't be able to tell whether you've got code, machine intelligence, or the real thing."

According to Chworktap, the beautiful female protein robot in Kilgore Trout's *Venus on the Half-Shell*, "Anything that has a brain complex enough to use language in a witty or creative manner has to have self-consciousness and free will." I don't know if I agree entirely with Chworktap. However, certainly within this century, some computers will respond in such a way that anyone interacting them will consider them conscious. The entities will exhibit emotions. Over time, we will merge with these creatures. We will become one. We will download our thoughts and memories to these devices. Our organs may fail and turn to dust, but our Elysian essences will survive.

Computers, or computer/human hybrids, will surpass humans in every area, from art to mathematics to music to sheer intellect. I do not know when this will happen, but I think it very likely that it will happen in this century. Of course, computers already exceed human "intelligence" when it comes to winning chess or solving certain mathematical problems. I see no reason this basic skill won't gradually metastasize into other areas like painting, music, and literature.

We will also become immortal in this century because we will fully understand the biological basis of aging. Already, scientists are able to knock out a gene called *daf-2*, or its equivalent, to make worms, flies, and mice live longer. One theory is that aging is programmed into organisms to reduce competition between old generations and younger. With this new understanding of aging and genes like *daf-2*, a number of colleagues believe that someone living in 2001 will be alive in 2150.

Personal Experience with Psychedelics

I have not had drug-induced psychedelic experiences, although many people who read my Neoreality books seem to think I have had chemically enhanced hallucinations. In some sense, I have to put a damper on my creativity and visions. Some of my publishers even tell me I write too much and should slow down. Nevertheless, my mind, visions, and ideas continually fly. As Salvador Dali said, "I am the drug!"

My Neoreality books discussed in Chapter 8 deal with Noah's Ark, sexy blonde supermodels with gibbon brains, fractal sex, cyanocobalamin, intellectual arthropods, and God in the form of moldy cheese. Is that sufficiently psychedelic for you? My artwork is also quite psychedelic and featured at such places as Erowid.org, alongside art produced by people under the influence of psychedelic drugs.

Extinction of the Human Species

Sometimes I wonder if the human species will survive the next hundred years. Could the human race destroy itself even if it wanted to? What about Doomsday Machines?

The term "Doomsday Machine" refers to the class of hypothetical weapons specially designed to destroy all large life forms including humans. Could such a weapon be produced? Sadly the answer may be yes, and several recipes have been given by Daniel Cohen, author of *Waiting for the Apocalypse.*

The easiest Doomsday Machine to construct is the cobalt bomb cluster. Each cobalt bomb is an ordinary atomic bomb encased in a jacket of cobalt. When a cobalt bomb explodes, it spreads a huge amount of radiation. If enough of these bombs were exploded, life on Earth would perish. In another recipe for Doomsday, large hydrogen bombs are placed at strategic locations on Earth and exploded simultaneously. As a result, the Earth may wobble on its axis. If placed at major fault lines, the bombs could trigger a worldwide series of killer earthquakes.

It may also be possible to capture one of the larger asteroids and send it crashing to Earth by exploding nuclear bombs at specific locations on the surface of the asteroid. Biological Doomsday Machines include weapons utilizing bacteria, viruses, or various biological toxins. For example, a few pounds of poison produced by botulism bacteria is sufficient to kill all human life.

Many believe that the Earth is like an inmate waiting on death row. Even if we do not die by a comet or asteroid impact, we know the Earth's days are numbered. The Earth's rotation is slowing down. Far in the future, day lengths will be longer than today's months. The Moon will hang in the same place in the sky, and the lunar tides will stop.

In five billion years, the fuel in our Sun will be exhausted, and the Sun will begin to die and expand, becoming a red giant. At some point, our oceans will boil away. No one on Earth will be alive to see a red glow filling most of the sky. As Freeman Dyson once said, "No matter how deep we burrow into the Earth...we can only postpone by a few million years our miserable end."

Today, our good friends at the "Voluntary Human Extinction Movement" (www.vhemt.org) believe that we should phase out the human race by "voluntarily ceasing to breed" to allow the Earth's biosphere to return to good health. I don't see this movement making a major impact. Do you? However, there exist many agents that might decimate our population: prions (infectious proteins); nanotechnology's gruesome, all-consuming gray goo (sub-micron-sized self-replicating robots, programmed to make copies of themselves, which get out of control, forming a thick, messy substance that coats the Earth); and terrorists' production of a biological agent like Ebola (an infectious virus).

On the other hand, I'm a bit more optimistic in the short run. Some researchers have even suggested that humans are at less risk for extinction now than at any other time in history, and that this risk decreases proportionately to advances made in technology. For example, aside from AIDS, it seems as if epidemics are less dangerous than in the days when the Europeans wiped out the South American Indians through disease and when Europe suffered from the Black Plague.

As I indicated, in this century we will probably become immortal from our understanding of the biological basis of aging and our merging with computers. Long before the Sun envelopes the Earth, we will have left it.

Rise of the Machine Civilization and Zygotic Personhood

Some readers have asked me if our own machines will one day rise up and destroy or dominate us. This is a common theme in many science-fiction stories. My response is always that it is more likely that we will blend with machines. We will become them. Similarly, we may one day be able to download ourselves to software and dispense with our physical bodies.

For the wealthy, genetic manipulation will cause us to become taller, more intelligent, more attractive, healthier, and stronger. For rich people, ugliness fades from the world. Our preoccupation with sexual pleasure will continue to increase in many segments of the population. One hundred years from now, humans will elect to have orgasms that last for hours, even days. With virtual reality, you'll be able to share the orgasm with whomever you want, from Paris Hilton and Brad Pitt, Marie Curie or Thomas Edison, to Alexander the Great, Queen Elizabeth, and Cleopatra.

In the future, stem cell research will become more common and help us evolve and keep healthy—despite the policies of people like George Bush, who prevented the federal funding of research to create stem-cell lines. On my RealityCarnival.com magazine, I wrote a letter to President Bush, asking him to keep an open mind and foster a liberal attitude with respect to a woman's option of having an abortion and with respect to embryonic stem cell research, which he had limited. The notion that an embryo or fertilized egg should be considered human is certainly open for debate. As reported in *Science* magazine, "zygotic personhood" (the idea that a fertilized egg is a person) is a recent concept.[20] For example, before 1869, the Catholic church believed that the embryo was not a person until it was 40 days old, at which time the soul entered. Aristotle also presumed this 40-day threshold. If the early embryo was soulless, perhaps early abortion was not murder. Pope Innocent III in 1211 determined that the time of ensoulment was anywhere from three to four months. In Jewish law, the fetus becomes a full-fledged human being when its head exits the womb. Before the embryo is 40 days old, it is "maya B'alma" or "mere water" (Talmud, Yevamoth 69b). If we truly believed that a zygote is a person, we would incarcerate women who use the pill because the pill sometimes prevents the implantation of a fertilized egg. We do not wish to jail such women or their physicians; hence, we do not actually believe a zygote is a person.

Birth-control methods have prevented the production of millions of unwanted children and lessened widespread suffering (including possible decreases in crime, child abuse, and ecological burdens). If people can overcome the fallacy of zygotic personhood, presidents can then ease restrictions on human embryonic stem cell research, which has the potential to help people with Parkinson's disease and diabetes. Although nonembryonic stem cells (such as multipotent adult progenitor cells) may eventually be suitable substitutes for embryonic cells, we should not restrict stem cell research now. Similarly, those who hope to ban cloning because it may entail the discarding of zygotes might rethink their position.

With women gaining more control over their reproductive fate, society has changed. Reliable birth control became as easy as taking a pill, which, along with education, is one of the greatest factors in helping women achieve equality with men and preventing overpopulation in less-developed parts of the world. Although religious people may debate whether a fertilized egg (zygote) should be accorded the same rights as a child (and therefore destruction of the zygote should lead to imprisonment), no one debates that the pill and other methods of birth control have decreased the suffering of fully formed, multicellular humans. Very few people today believe in gametic personhood (the idea that sperm and eggs are people) or homuncular personhood (the 18th-century idea that the entire human organism—the homunculus—is contained in the spermatozoa); similarly, the notion of zygotic personhood may someday fade from the world scene. This leads me to one of the more mind-boggling issues that we will face in the next century: the notion of cybernetic personhood.

In the coming years, we will be able to create sentient creatures in software running on computers. We will be able to simulate ourselves in software. This, of course, will affect laws, politics, and religion. The termination of sentient software may one day be much more egregious than termination of a zygote. Returning our attention to present technology, I hope that future U.S. presidents consider the appointment of individuals—both to the judiciary and to positions of policy making—who have not taken extreme positions in opposition to abortion or embryonic stem cell research.

What does the future hold? Many colleagues believe that we'll enhance our senses using genetic engineering. Within this century, some of us will extend our visual and auditory ranges and have synesthesial senses we can barely imagine. We'll be in constant contact with one another through wireless mindlinks.

At first, we'll see the nascent seeds of these mindlinks in the form of implantable cell phones. Shortly thereafter, we'll become more sophisticated. Already, technologists are creating vocoders that convert nerve signals in the vocal chords to computerized speech. Cochlear implants convert sounds into neural signals that the brain can interpret. By interfacing the vocoder and cochlear implant with radio transmitters, we can take the first steps to e-telepathy, kiss the acoustic age good-bye, and enter the realm of thought-to-thought communication.

Cell phones and e-mail began to transform the planet around the year 2000. Imagine the transformative potential of e-telepathy in the next fifty years. Scientific, artistic, and political collaborations that took months a hundred years ago, could be done in a flash.

Musing about these kinds of direct mindlinks, George Dvorsky of BetterHumans.Com notes

On the surface humanity appears to be spreading outward, venturing across continents and into space. Yet in actuality, we are journeying towards one another. Our globe has never appeared smaller and our proximity to each other has never been closer. This trend shows no signs of slowing down, pointing the way to a remarkable interconnected future.[21]

Within 15 years, stopwatch-sized Vagus Nerve Stimulators (VNSs) will be prevalent as a means of making us feel happier. (Temple University's Jake Zabara showed over a decade ago that the VNS can stop epileptic seizures, and today we believe the VNS can also cure depression by zapping the vagus nerve in the neck.[22]) Several research teams in the US and in Europe are already engineering new varieties of mosquitos whose bite actually prevents malaria and other diseases by injecting antibacterial toxins.[23] Other scientists are designing silkworms with DNA from humans so that the worms spin proteins like collagen that have pharmaceutical and industrial uses.[24] Within 20 years, genetically modified creepy crawlers will push humanity into the Superbug Age in which insects inexpensively create a limitless supply of novel materials for our buildings and bodies. Right now, genetically modified tilapia fish are churning out lifesaving human blood-clotting factors.[25] The fish, an early symbol of Christianity, will be our saviors.

If intelligent aliens exist, perhaps we will one day discover a message from extraterrestrials in our own genetic code. A chapter in my book *Mazes for the Mind* describes the hypothetical discovery of the digits of pi encoded in the DNA of a tarsier (a small mammal), and the incredible impact such a discovery would have on humanity. We can imagine several ways in which pi could be encoded in DNA, for example, using base-4 arithmetic and assigning values to the DNA bases ($G = 0$, $C = 1$, $A = 2$, and $T = 3$). The number 314159 in base 10 would be represented as 1030230233 in base 4, where no digit can be greater than 3. Therefore, CGTGATGATT would code for the first 6 digits of pi. Certain repeating sequences can be used to coax the biologist to pay attention to a subsequent alien message in the DNA string. Aliens could also code a number using repeated DNA bases followed by CAT as a delimiter. Using this notation, we can avoid a dependence on base 10 numbers, and a list of the first few prime numbers, 2, 3, 5, 7, 11, 13, would be:

GGCATGGGCATGGGGGCATGGGGGGGGCAT
GGGGGGGGGGGGGCATGGGGGGGGGGGGGGGGGCAT

In *Mazes for the Mind*, the discovery of the tarsier DNA message eventually led to a ban on all trade in tarsiers, because scientists, in their zeal to study the tarsier in greater detail, depleted the world population of these mammals! Scientists then rushed madly to sequence and study other similar tree-dwellers such as the slender loris, a lemur from southern India, in search for more messages from the cosmos. Chaos ensues.

We can expect startling demographic changes in world populations during this century and the next. Some say that increased racial mixing will continue to take place, and that the last blonde will die in Finland in 400 years. This is probably not literally true, but I do believe that many peoples and some religions are dying out. As I discussed in Chapter 1, because more than 50 percent of Jews in America are marrying people from other religions, the last practicing non-Orthodox Jew will die in America in 300 years. According to Alan Dershowitz, a Harvard study predicts that the American Jewish community is likely to number less than 1 million and conceivably as few as 10,000 by the time the United States celebrates its tricentennial in 2076. Jews may only consist of isolated pockets of ultra-Orthodox Hasidim.

Sometimes I mourn because the ultimate fate of the universe involves great cold—or great heat if there's sufficient gravity to draw all matter together in a single point in a final Big Crunch. It is likely that *Homo sapiens* will become extinct. However, our civilization and our values may not be doomed. Our silicic heirs, whatever or whomever they may be, may find practical ways for manipulating spacetime as they launch themselves throughout the galaxy. They will seek their salvation as sentient simulacra in the stars.

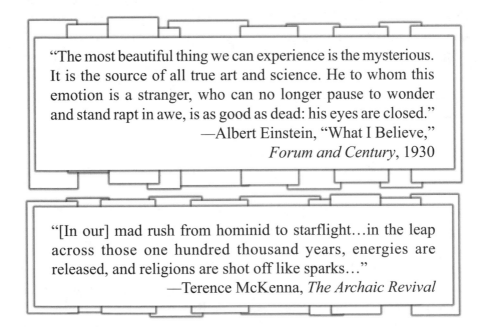

"The most beautiful thing we can experience is the mysterious. It is the source of all true art and science. He to whom this emotion is a stranger, who can no longer pause to wonder and stand rapt in awe, is as good as dead: his eyes are closed."
—Albert Einstein, "What I Believe,"
Forum and Century, 1930

"[In our] mad rush from hominid to starflight...in the leap across those one hundred thousand years, energies are released, and religions are shot off like sparks..."
—Terence McKenna, *The Archaic Revival*

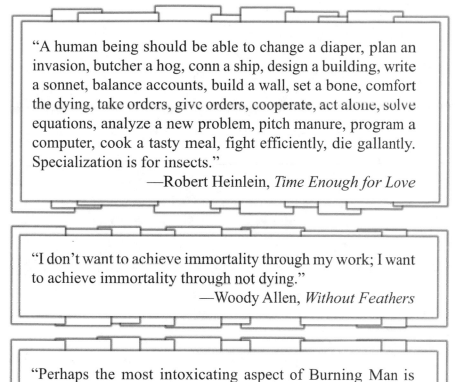

"A human being should be able to change a diaper, plan an invasion, butcher a hog, conn a ship, design a building, write a sonnet, balance accounts, build a wall, set a bone, comfort the dying, take orders, give orders, cooperate, act alone, solve equations, analyze a new problem, pitch manure, program a computer, cook a tasty meal, fight efficiently, die gallantly. Specialization is for insects."
—Robert Heinlein, *Time Enough for Love*

"I don't want to achieve immortality through my work; I want to achieve immortality through not dying."
—Woody Allen, *Without Feathers*

"Perhaps the most intoxicating aspect of Burning Man is thinking about Burning Man."
—Daniel Pinchbeck, *Breaking Open the Head*

Farewell

Good-bye, Proust.

In Chapter 6, we mentioned two magnificent walking paths that author Marcel Proust followed on his daily walks through the town of Combray, France. Reminiscing about the paths filled his mind with joy and fantasy, just as I feel with my own walks through Shrub Oak. Proust recognized that events in his life were attached to his two paths like little trinkets dangling from the band of a charm bracelet.

In *Time Regained*, the last volume of Proust's novel, he discovers that the Meseglise way and Guermantes way are joined, and are actually the *same* path. Two paths, seemingly very different in childhood, are united as he grows old.

Every now and then, when I fall asleep, I imagine myself choosing between Proust's Meseglise way and the Guermantes way. The Meseglise way had a gorgeous view of the plains as it meandered through lilacs, hawthorns, cornflowers, poppies, and apple trees. The Guermantes way traversed a landscape of rivers and ponds filled with tadpoles and water-lilies. Either way led to interesting people, beauty, and adventure. In *In Search of Lost Time*, we see "fine greenery of the park and the lilacs at the entrance, over the green leaves of the tall trees by the edge of the lake, sparkling in the sun, and the forest of Meseglise." Proust said that the Guermantes way represented a place in which he could pass his whole life—his only requirements were that he could "go out fishing, drift idly in a boat, see the ruins of a gothic fortress in the grass, and find hidden among the cornfields an old church…"[1] For Proust, no other rivers were as fair as those by the Guermantes way.

Marcel's father in *In Search of Lost Time* took his family on many walks through the meadows and fields of Combray. In some scenes, the sun would descend as twilight enveloped the world. Little puffs of vapor hovered over the grasses, and crickets chirped. Father would take his loving family down new paths and unfamiliar trails, until the family was thoroughly lost. And then they'd look more closely and realized they were standing outside the garden gate of their own home. Ursula LeGuin in The *Dispossessed* said, "True journey is return." John Fowles in *The Magus* captured the essence of Proust, and the foundation of the world, in a single statement:

> I looked round the trees. The thin net of reality. These trees, this sun. I was infinitely far from home. The profoundest distances are never geographical.

Proust, like some post-exilic prophet, lived in a spirit world sandwiched between this world and the world beyond, where even a pattern on the wallpaper became fodder for hyper-speculation and dreams: "For it was old wallpaper on which every rose was so distinct that, had it been alive, you could have picked it, every bird you could have put in a cage and tamed, quite different from those grandiose bedroom decorations of today where, on a silver background, all the apple trees of Normandy display their outlines in the Japanese style to hallucinate the hours you spend in bed…"

A recent survey conducted in England revealed that Marcel Proust is the most influential and admired novelist. Moreover, he was thought to have the most enduring influence on the 21st century of any other writer.[2] Proust biographer Edmund White says that Proust's fame and prestige eclipses those of James Joyce, Samuel Beckett, Virginia Woolf, William Faulker, Ernest Hemingway, Thomas Mann, and F. Scott Fitzgerald. Harvard graduate and renowned contemporary novelist Andrew Holleran once remarked that *In Search of Lost Time* was so perfect that it "ended the novel simply by doing something so complete, monumental, perfect, that what the fuck can you do afterwards?" Novelist Jean Genet read the opening pages of Proust and then closed the book, hoping to savor every sentence, every thought, every metaphor, every image. Genet wanted the reading experience to linger for as long as possible, and said to himself, "Now, I'm tranquil. I know I'm going to go from marvel to marvel."[3]

Perhaps that's what life should be: a slow trip from marvel to marvel. We pass through the lives of so many people, barely slowing down and enjoying their presences. We spend too much time in the past, yearning to right some wrong or longing for more youthful days. We also spend too much time thinking about the future, without enjoying the "now."

The characters in Proust's novel morphed and shimmered like ghosts or the *doppelgängers* seen by people with the strange psychiatric disorders we discussed in Chapter 5. In *In Search of Lost Time*, the narrator's lover Albertine is mentioned 2,360 times, but the precise location of her beauty mark continues to change from her lip to her chin to below her eye.[4] Some of his characters, like the group of girls in Cabourg (a seaside resort in Normandy that became the fictional Balbec of his novel), form a hive mind or swarm creature rather than act as individuals. The band of girls roam through the beach town like a collection of particles controlled by a computer program that simulates the flocking of birds.

I often wonder if Proust's frail health contributed to his passion for writing *In Search of Lost Time*. Sometimes, he would have a dozen asthma attacks a day, which left him extremely weak. Through history, geniuses like Proust differed from the norm not only in intelligence, but also in some other mental or physical characteristic as well. For example, Alexander, Aristotle, Archimedes, Attila, John Hunter, William Blake, and St. Francis Xavier were all short (most less than 5'2"). Many great people have had a deformity of one kind of another: Susan B. Anthony (crossed eyes), Vladislav Khodasevich (six fingers), Jane Addams (spinal disfigurment), Ring Lardner (deformed foot), Allen Dulles (club foot), Claude Debussy (large, bony protuberances on forehead). Aristotle, Aesop, Demosthenes, Virgil, Darwin, and Cavendish all stammered.[5] Many geniuses had asthma, incapacitating allergies, obsessive-compulsive disorder, or bipolar disorder. Could it be that a distinctive physical appearance or chronic physical ailments give some individuals the desire to compensate for their shortcomings, or to leave a mark on the world and achieve immortality through creative excellence?

Proust had plenty of physical problems. Even when he was 38, Proust worried that he would soon die. His favorite quote from the Bible was "Walk while ye have the light" (John 12:35). Over the next decade, he used this imperative as a stimulus to create.

And what finally became of Proust? He continued to write in his bed, even when he was close to death. In fact, he used the experience of dying as material for the death of one of characters, Bergotte. Milton L. Miller writes in *Nostalgia: A Psychoanalytic Study of Marcel Proust*, "It was on the morning after November 17, when he believed himself much better, and kept his brother near him a long while, that, at about 3 a.m., he called

Céleste, and although choking and feeling worse again, he dictated supplementary notes about Bergotte's death....At the very last, there were indecipherable scribblings, in which he tried to write something about Forcheville."

Time Regained, the final volume of Proust's *In Search of Lost Time*, revolves around aging, illness, and death. The events of *Time Regained* take place after the First World War. The narrator "Marcel" has been ill, but recovers and returns to Paris to attend an opulent society party. Marcel seems to have entered a parallel universe in which friends and acquaintances from the past are wearing costumes and posing as old men or women. But gradually, Marcel realizes that the people's strange appearance results from their *actual* aging. Time has passed. Marcel's memories are decades old. Many of his friends have died and are barely remembered. Ghosts are everywhere as Proust fades from the realm of man.

As I finish writing the closing words of this book, I imagine myself walking along an infinite Guermantes way, dissected from reality with a mile on each side of the path and placed in outer space. That's all I need to capture the dream—a mere mile of scenery on each side of the path. I dream in particular of the Guermantes way, because it was so long that Proust never got to its end. For Proust, the Guermantes was a goal, an abstract geographical entity like the North Pole or the Equator. During the course of his mystic walks, Proust never found the source of the nearby, crystal-clear Vivonne River, although he continually yearned to find its beginning. He could never reach his goal of attaining the Guermantes estate itself. Despite the impossible vistas, he continued to yearn for long travels.

Come with me for a moment. Imagine we are now walking along Proust's other favorite path, the Meseglise way. In the distance is the fringe of the Roussainville woods with its dense thatch of leaves. Imagine that we are with the spirit of Proust as he observes the universe around him. From his novel: "Out of the fresh little green hearts of their foliage, the lilacs raised inquisitively over the fence of the park their plumes of white or purple blossom, which glowed, even in the shade, with the sunlight in which they had been bathed."[6]

Let's walk little further. We pause by a wooden fence, gazing at the underbrush, and Proust speaks his dreamlike prose:

Lilac-time was nearly over; some of the trees still thrust aloft, in tall purple chandeliers, their tiny balls of blossom, but in many places among their foliage where, only a week before, they had still been breaking in waves of fragrant foam, these were now spent and shriveled and discoloured....dry and scentless....[7]

We pass "two tiers woven of trailing forget-me-nots below and of periwinkle flowers above." A little further and we see "delicate, blue garland which binds the luminous, shadowed brows of water-nymphs; while the iris, its swords sweeping every way in regal profusion, stretched out over agrimony and water-growing king-cups the lilied sceptres, tattered glories of yellow and purple, of the kingdom of the lake."[8]

Are you catching the essence of Proust, as my book comes to an end? Isn't he better than a psychedelic trip? Do you think he saw the same universe that we do, or did he reside in a world beyond our own? Dream deep, as Proust rambles on:

I found the whole path throbbing with the fragrance of hawthorn-blossom. The hedge resembled a series of chapels, whose walls were no longer visible under the mountains of flowers that were heaped upon their altars...[The flowers] held out each its little bunch of glittering stamens with an air of inattention, fine, radiating "nerves" in the flamboyant style of architecture, like those which, in church, framed the stair to the rood-loft or closed the perpendicular tracery of the windows, but here spread out into pools of fleshy white, like strawberry-beds in spring.[9]

Go deeper:

My eyes followed up the slope which, outside the hedge, rose steeply to the fields, a poppy that had strayed and been lost by its fellows...hoisting upon its slender rigging and holding against the breeze its scarlet ensign, over the buoy of rich black earth from which it sprang, made my heart beat as does a wayfarer's when he perceives, upon some low-lying ground, an old and broken boat which is being caulked and made seaworthy, and cries out, although he has not yet caught sight of it, "The Sea!"[10]

Go one level deeper, and then you can brag to all your friends that you have actually read Proust, have experienced Proust in his fullest:

> And then I returned to my hawthorns, and stood before them as one stands before those masterpieces of painting which, one imagines, one will be better able to "take in" when one has looked away, for a moment, at something else; but in vain did I shape my fingers into a frame, so as to have nothing but the hawthorns before my eyes; the sentiment which they aroused in me remained obscure and vague, struggling and failing to free itself, to float across and become one with the flowers. They themselves offered me no enlightenment, and I could not call upon any other flowers to satisfy this mysterious longing.[11]

I sometimes can shape my dreams and force content into them. Tonight I will float along the Vivonne River by the Guermantes way. I visualize it now. My tranquil life seems like an afterlife or a dream—unreal. My friends who try this mental exercise come to believe that life can be repeated, refined, and relived with slight alterations and sadness avoided. All this lovely play of form and light and color on the Vivonne River and in the eyes of humans is no more than that: a playing of illusions in spacetime.

"Transcend," I whisper, as I trail my fingers in the clear fluid of the Vivonne. Within its depths, a fiery orange light reflects off spirals on tessellated water, challenging the azure of an endless fractal sky. I smell lilacs, and then....

My little boat turns and glides above bright shining swirls of mist and a symmetrical melange of intertwined, golden spirals. The elves chuckle. I am moving toward open sea.

> "Doubtless my books also, like my earthly being, would finally some day die. But one must resign oneself to the idea of death. One accepts the idea that in ten years one's self, and in a hundred years one's books, will no longer exist. Eternal existence is not promised to books any more than to men."
> —Marcel Proust, *The Past Recaptured*[12]

"What if our universe started out as not quite real, a sort of illusion, as the Hindu religion teaches, and God, out of love and kindness for us, is slowly transmuting it, slowly *and secretly*, into something real?"
—Philip K. Dick, "How to Build a Universe That Doesn't Fall Apart Two Days Later," in *I Hope I Shall Arrive Soon*

"Physical concepts are free creations of the human mind, and are not, however it may seem, uniquely determined by the external world. In our endeavor to understand reality, we are somewhat like a man trying to understand the mechanism of a closed watch. He sees the face and the moving hands, even hears it ticking, but he has no way of opening the case. If he is ingenious, he may form some picture of the mechanism which could be responsible for all the things he observes, but he may never be quite sure his picture is the only one which could explain his observations. He will never be able to compare his picture with the real mechanism, and he cannot even imagine the possibility of the meaning of such a comparison."
—Albert Einstein, *The Evolution of Physics*, 1938

"When, on a summer evening, the resounding sky growls like a tawny lion, and everyone is complaining of the storm, it is along the 'Méséglise way' that my fancy strays alone in ecstasy, inhaling, through the noise of falling rain, the odour of invisible and persistent lilac-trees."
—Marcel Proust, *Swann's Way*[13]

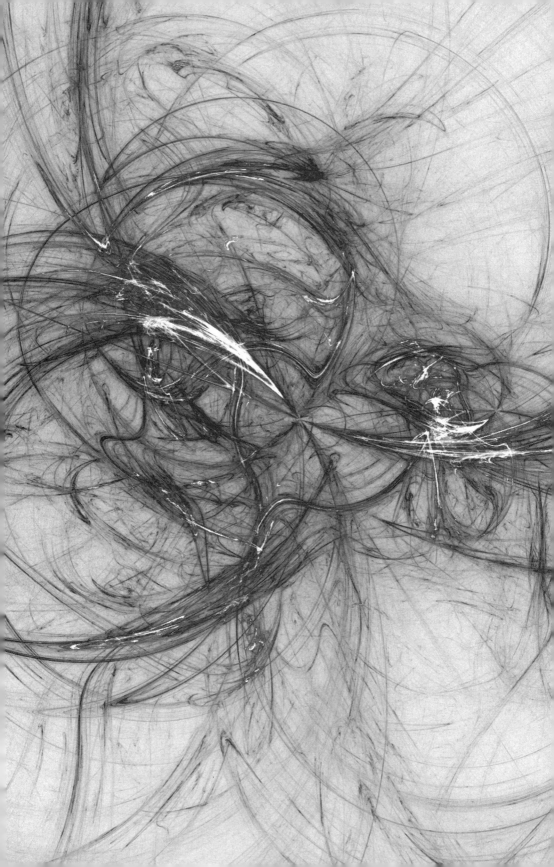

Epilogue

"Sleep is one-sixtieth a part of death. A dream is one-sixtieth part of prophecy."

—Talmud, Berachot 57b

I write these words at the end of my Mediterranean quest that I described in the Preface. I'm at the topmost point of the medieval village of Eze in France, overlooking the French Riviera. The scent of jasmine fills the air, and the carmine colors of bougainvillea flood my eyes. I touch the remains of the stone walls around me that date back to the Bronze Age.

My feet are tired, but my spirit soars. I've seen the "head of Christ" in Michelangelo's Pieta in the Vatican and gazed into the eyes of the Black Madonna in Montserrat. I've heard the different languages, learned new words, and saw the dreamy cathedral towers of Antoni Gaudí in Barcelona. Although I never gazed at Einstein's or Proust's brain, at least while in Florence I could kneel at the bones of Galileo, one of Einstein's greatest scientific predecessors, and Rossini, the genius Italian musical composer who wrote more than 30 operas with lightning speed.

Now, in Eze, the remains of the castle reach toward heaven like the ribs of some giant creature. The soldiers of Louis XIVth destroyed much of the fortress in 1706. Yet the bougainvillea remain, healthy, strong. I imagine their branches writhing and finally coalescing, forming the motto of Eze, "*Moriendo Renascor*"—"In death I am reborn."

Feedback

> "I like to build universes which do fall apart. I like to see them come unglued…"
> —Philip K. Dick, "How to Build a Universe That Doesn't Fall Apart Two Days Later," in *I Hope I Shall Arrive Soon*

In the following pages, I've compiled a list of notes as well as further reading that reference, discuss, and identify much of the material I used to research and write this book. I include Internet Web sites in addition to books and journals. As many readers are aware, Internet Web sites come and go. Sometimes they change addresses or completely disappear. The Web site addresses listed here provided valuable background information when this book was written. You can, of course, find numerous other Web sites relating to the curiosities discussed in this book by using Web search tools such as the ones provided at www.google.com.

If I have overlooked an interesting cultural curiosity, person, reference, or factoid—which you feel has never been fully appreciated—please let me know about it. Just visit my Web site www.pickover.com, and send me an e-mail explaining the idea and how you feel it influences our perception of reality.

Future books written in the spirit of *Sex, Drugs, Einstein, and Elves* may include topics listed in the following paragraphs. Please let me know which of these topics interests you most, and add your own topics to the list. The topics receiving the most votes will be included in future books:

Transubstantiation, the metaphysics of Melungeons, synchronicity, Button Gwinnett, arianism, panpsychism, the Ark of the Covenant, communication with the dead, Chuck Palahniuk, *Andromeda* (the TV Show), the *Bakhshali* manuscript, wikipedia.org, crosses of light, judaizing Sabbatarianism, the Quadratrix of Hippias, James Joyce, demons, feral children, hollow earth, a neo-Kemetic interpretation of Ancient Egypt, *Jugendstil*, Ivy League nude posture photos, Saint-Simonianism and the doctrine of hope, Mary Celeste, Chédiak-Higashi syndrome, Rolf Tiedemann, Jewish teachings on reincarnation, *The Exorcist*.

Charcot-Wilbrand syndrome and the brain's dream cinema, Noah's ark, the planet Vulcan, the monochromatic Muslim chessboards, the powers of holy water, relics of the true cross, Borromean rings, savant syndrome, stigmata, subliminal messages, teleportation, Truman Capote's *Music for Chameleons*, aliens, Amazon.com, Walt Disney, the Antichrist, Adramelech, Arthur C. Clarke, ekpyrotic creation of the universe, the Seven Valleys of Bahá'u'lláh, Soka Gakkai, Carl Sagan, cellular automata, the ninth wave, gualac wood oil, dolphins, ESP, the fourth dimension, Mr. Ub Iwerks, Nazca lines, fractals, Freeman Dyson, the ecstasy of Hasidism, *Nannophya pygmaea*.

Brewster Higley, Google, hoaxes, hyperspace, immortality, infinity, IQ, irrational numbers, Isaac Asimov, kartoo.com, kung fu, lucid dreaming, palindromes, paradoxes, the Piri Reis map, Lemuria, parallel universes, The Grateful Dead's telepathic adventure, primates, prime numbers, the Copeland-Erdös constant, prodromic dreams, Andrealphus, the baculum of bats, quantum physics, Robert Heinlein, rains of mcat and other effluvia, secret codes, Khoomi throat singers of Mongolia, Thomas Pynchon, Mongolian death worms, time travel, transcendental numbers, the Sedlec Ossuary, *Eozoon canadense* (the "false dawn" animal), transhumanism, virtual reality, string theory, Tetzkatlipoka and the Aztec tradition of consciousness, Cantor's infinity.

The Baal Shem Tov, annelid worms that turn skin cells into brain cells, the mystery of IYNKICIDU, the Loretto miracle staircase, Merovingians, Goan of Vilna, Hans Jonas, Shabbetai Zvi, the bridegroom of blood, Antikythera Mechanism, nephilum, indigo children, David Icke, Cocoa Puffs, the Shi'ur Komah's dimensions of God, the mass suicide of the Xhosa, Nongqawuse—prophetess of doom, the Baghdad Battery, *The Da Vinci Code*, Xanthian marbles, Chebyshev polynomials, Azazel, Dairy Queen, Yggdrasill, the super-Sargasso Sea in our atmosphere, Nawi Ollin Teotl and the Mexhika system of decision making, angel hair (believers say it is a manifestation from a higher dimension; skeptics say it is a cobweblike amorphous mass of calcium, silicon, magnesium and boron, spawned from the atmosphere in some unknown manner).

"Proust is the great poet of memory but he achieves this by being such a great master of metaphor and by showing the way that every world (the inner landscape of any mind) contains a thousand worlds....In *Remembrance of Things Past* he has produced that rarest of all things, a book of wisdom. This is a book like the Bible or the works of Shakespeare that a person can live in. On any desert island, in any prison cell, if you have Proust beside you, your life will be richer."
—The Age Company, "Give Proust a Chance"

Notes

"If only acid [LSD] were legal, I think I would recommend that some patients do it."
—Andrew Weil, M.D., *60 Minutes*, March 2001 (One of the biggest names in alternative medicine, Dr. Weil is alleged to have tried drugs ranging from LSD to toad venom and Yage.)

Dedication

1. Dyson, Freeman, "Introduction." In Cornwell, John (editor), *Nature's Imagination: The Frontiers of Scientific Vision* (New York: Oxford University Press, 2000).

Table of Contents

1. Clifford A. Pickover.
2. Clifford A. Pickover.
3. Clifford A. Pickover.
4. Clifford A. Pickover.
5. Clifford A. Pickover.
6. Clifford A. Pickover.
7. Ancient Chinese proverb.
8. Rumi (1207–1273), Sufi poet and mystic. Jalal al-Din Muhammad Rumi was born in the region today known as Afghanistan. His family fled the Mogul invasion to Konya, Turkey, where he spent most of his life.
9. Marian Diamond, *OMNI* magazine, c. 1995. When I asked Diamond if she knew the precise date, she responded: "I have used this sentence many times in public speeches (some 800+ in all) and have no idea where it has been printed." Diamond is one of the world's great neuroscientists, a writer, and a compelling lecturer. She's Professor of Neuroanatomy at the University of California at Berkeley.

10. Clifford A. Pickover.

Preface

1. Exodus 35: 30–36:1. This particular English translation of the Torah is at "Chabad Online Network, A Division of the Chabad-Lubavitch Media Center," http://www.chabad.org/parshah/rashi/text.asp? MosadTitle2=Chabad%2Eorg&AID=15569&p=2. See also the Judaica Press translation, "The Judaica Press Complete Tanach with Rashi," http://www.judaicapress.com/tanach_with_rashi.asp Readers may enjoy consulting http://bible.gospelcom.net/ to view many different translations of this passage.

Introduction

1. I amplify these feelings in Pickover, Clifford, *The Paradox of God* (New York: St. Martin's Press/Palgrave, 2002).

2. Kanigel, Robert, *The Man Who Knew Infinity* (New York: Scribner, 1991). See Pickover, Clifford, *A Passion for Mathematics* (New York: John Wiley & Sons, 2006), in which I also discuss Ramanujan and other creative people who have been open to dreams as a source of inspiration.

3. Abraham, Ralph, in a 1991 interview with *GQ* magazine. Quoted in Kick, Russ, *The Disinformation Book of Lists* (New York: The Disinformation Company, 2004), 14.

4. James, William, *The Varieties of Religious Experience: A Study in Human Nature* (New York: Modern Library; reprint edition, 1994), first published in 1902.

5. Ibid.

6. Dinitia Smith, "Why Proust? And Why Now?" *The New York Times*, April 13, 2000, http://www.nytimes.com/library/books/041300 proust.html

7. Proust, Marcel (author), with translations by Andreas Mayor and Terence Kilmartin, *In Search of Lost Time Volume VI, Time Regained* (New York: Modern Library, 1999). See chapter 3, "An Afternoon Party at the House of the Princesse de Guermantes."

8. *The Sopranos*, (HBO TV show), Season Three: "Fortunate Son," Scene 1.

9. Paul Ehrenfest initiated a famous physics colloquium series in 1912 where he encouraged students to ask questions, and he said that no questions werc "stupid."

10. Murphy, Michael, *The Future of the Body: Explorations into the Further Evolution of Human Nature* (New York: Putnam, 1992).

11. 1929 interview with German-born American interviewer George Sylvester Viereck. In addition to being a consummate interviewer himself, Viereck was a celebrated poet, political writer, and editor. For details on the interview setting, see Brian, Denis, *Einstein: A Life* (New York: John Wiley & Sons, 1996), 185.

12. Steinbeck, John and Edward Flanders Ricketts, *The Log from the Sea of Cortez* (New York: Penguin Books; reprint edition, 1995).

Chapter 1
On Fugu Sushi and Transdimensional Reality Worms

1. Clifford A. Pickover.

2. Constant, Lawrence, "Commentary: The History of Shrub Oak," *North County News*, Vol. 3, No. 34, January 28–February 3, 1981, http://www.yorktownhistory.org/ncn/places/shrub_oak/history_of_shrub_oak.htm (Lawrence Constant is the pen name of a local historian.)

3. I found a number of Proust quotes, such as this, in de Botton, Alain, *How Proust Can Change Your Life* (New York: Pantheon, 1997).

4. Scovell, Nell, "One Fish, Two Fish, Blowfish, Blue Fish," *The Simpsons* TV show. Episode Number 24, Production Code 7F11, First Aired January 24, 1991.

5. Britton, Everard B. "A Pointer to a New Hallucinogen of Insect Origin," *Journal of Ethnopharmacology*, 12(3): 331–333, December 1984.

6. Barr, Cameron W., "Cicada: The Other, Other White Meat: Epicures Ready to Make a Meal of High-Pitched Pests." See http://www.msnbc.msn.com/id/4752983/ (April 16, 2004), for reprint of a story from the *The Washington Post*. (This article discusses Grubco, Inc.)

7. Kick, Russ, *The Disinformation Book of Lists* (New York: The Disinformation Company, 2004), 44. See also Rudgley, Richard, *The Encyclopedia of Psychoactive Substances* (New York: St. Martin's Press, 2000).

8. Ibid.

9. Rudgley, Richard, *The Encyclopedia of Psychoactive Substances* (New York: St. Martin's Press, 2000).

10. Horgan, John, *Rational Mysticism* (New York: Houghton Mifflin, 2003).

11. Samorini, Giorgio, *Animals and Psychedelics* (Rochester, Vermont: Park Street Press, 2000).

12. Davis, Erik, "Calling Cthulhu: H. P. Lovecraft's Magick Realism." In Metzger, Richard, *Book of Lies: The Disinformation Guide to Magick and the Occult* (New York: The Disinformation Company, 2003), 138–139.

13. Lovecraft, H. P., "From Beyond," written in 1920, first published in *The Fantasy Fan*, June 1934. The full text is at many Web sites.

14. Ibid.

15. James, William, *The Varieties of Religious Experience: A Study in Human Nature* (New York: Modern Library; reprint edition, 1994).

16. *Contact* (movie), starring Jodie Foster and Matthew McConaughey. Directed by Robert Zemeckis, Warner Studios, 1997. Based on Carl Sagan's 1985 novel.

17. This quotation comes from an anonymous essay that Sagan wrote for Grinspoon, Lester, *Marijuana Reconsidered* (Oakland, California: Quick American Archives; 2nd edition, 1994). Also see Poundstone, William, *Carl Sagan: A Life in the Cosmos* (New York: Henry Holt & Company, 1999); and Kick, Russ, *The Disinformation Book of Lists* (New York: The Disinformation Company, 2004), 21.

Chapter 2
The Quantum Mechanics of Hopi Indians

1. Clifford A. Pickover.

2. Webster, Diane, *Getting to Know Yorktown* (Brewster, New York: Guide Communications, 2004). Also see Cooper, Linda and Alice Roker, *Images of America: Yorktown* (New York: Arcadia Publishing, 2003).

3. Dyson, Freeman, "The Two Windows." In Templeton, John Marks (editor), *How Large is God?* (Radnor, Pennyslvania: Templeton Foundation Press, 1997).

4. Medlej, Joumana, "Knock, Knock," http://www.cedarseed.com/ and in particular, see "Thinking in Tongues," http://www.cedarseed.com/air/blabla3.html

5. Ibid. According to *New Scientist*, Koreans use a different verb to describe placing something into a container such that there is a loose fit or a tight fit. English speakers don't differentiate these two categories so easily, and, more generally, do not divide loose-fitting or tight-fitting objects into two clear groups as Korean speakers do. For more informa-

tion, see: Philips, Helen, "The Concepts Are There Even If the Words Aren't," *New Scientist*, 183(2457): 8, July 24, 2004.

6. Woodbury, Anthony C., "Counting Eskimo Words for Snow: A Citizen's Guide," University of Texas at Austin, July 1991, *LINGUIST List*: Volume 5-1239. (Discusses lexemes referring to snow and snow-related notions in Steven A. Jacobson's *Yup'ik Eskimo Dictionary*.) http://www.princeton.cdu/~browning/snow.html See also Professor Anthony Woodbury's Web site at the Department of Linguistics, University of Texas: http://www.utexas.edu/cola/depts/linguistics/

7. Ross, Philip, "Draining the Language Out of Color," *Scientific American*, 290(4): 46–47, April 2004. (Discusses the work of linguist Paul Kay.)

8. Ibid.

9. Strauss, Stephen, "Life without numbers in a unique Amazon tribe. Piraha Apparently Can't Learn to Count and Have No Distinct Words for Colours," *The Globe and Mail*, August 20, 2004, Page A3, http://www.theglobeandmail.com/scrvlet/ArticleNews/TPStory/LAC/20040820/NUMBERS20/TPScience/. See also, Gordon, Peter, "Numerical Cognition Without Words: Evidence from Amazonia," *Science*, August, 19 2004, http://www.sciencemag.org/cgi/content/abstract/1094492v1 and also, Holden, Constance, "Life Without Numbers in the Amazon," *Science,* 305(5687): 1093, August 20, 2004.

10. Ibid.

11. Ibid.

12. Science Daily, "Study of Obscure Amazon Tribe Sheds New Light on How Language Affects Perception," adapted from a news release issued by Columbia University Teachers College, August 20, 2004, http://www.sciencedaily.com/releases/2004/08/040820083420.htm

13. Ibid.

14. Ibid.

15. Brewer, Ebenezer Cobham, *Brewer's Dictionary of Phrase and Fable* (New York: HarperResource; 15th edition, 1995; first published in 1870).

16. Ibid.

17. Einstein said this to psychologist Max Wertheimer, as recounted in Wertheimer, Max, *Productive Thinking* (New York: Harper, 1959), 213–228.

18. Medlej, Joumana, "Thinking in Tongues," http://www.cedarseed.com/air/blabla3.html

19. Ibid.

20. Pinker, Steve, *The Language Instinct (New York: Harper*Perennial, 2000).

21. Malotki, Ekkehart, *Hopi Time: A Linguistic Analysis of the Temporal Concepts in the Hopi Language (Trends in Linguistics, Studies and Monographs, No. 20)* (Berlin: Walter De Gruyter, 1983).

22. Alford, Dan Moonhaw, "Whorf Hypothesis Hoax: Sin, Suffering and Redemption in Academe," Chapter Seven in *The Secret Life of Language*, October 17, 2002 draft. http://www.enformy.com/dma-Chap7.htm

23. An excellent review of *Do Animals Think* can be found in Clayton, Nicola, "An Open Sandwich or an Open Question?" *Science* 305(5682): 344, July 18, 2004. Here, Nicola discusses the work of Clive Wynne, Euan Macphail, and Jerry Fodor.

24. Grandin, Temple, *Thinking in Pictures: And Other Reports from My Life with Autism* (New York: Vintage Books, 1996). From Chapter 1, "Autism and Visual Thought."

25. Adams, Douglas, *The Restaurant at the End of the Universe* (New York: Ballantine Books; reprint edition, 1995).

26. Weyl, Herman, *Philosophy of Mathematics and Natural Science* (New York: Atheneum, 1963), 116.

27. In short, the Sapir-Whorf Hypothesis suggests that the structure of a language influences the way a person behaves and thinks, and it's a hypothesis many cognitive scientists, including Noam Chomsky and Steven Pinker, reject. Although the strong version of the basic hypothesis of linguistic determinism (language determining thought) is simplistic, words and concepts of our language surely have some influence on how we perform and perceive the world. If you don't have the words to describe your thoughts, you will not be able to convey your thoughts to others.

28. Sapir, Edward, *Language: An Introduction to the Study of Speech* (New York: Harvest Books; 1955).

29. Carroll, John B., *Language, Thought, and Reality: Selected Writings* (Cambridge, Massachusetts: MIT Press; 1964). Reprints of Benjamin Lee Whorf's writings.

30. "Babel's Children," *The Economist*, 370(8357): 61, January 8, 2004,http://www.economist.com/printedition/displayStory.cfm?

Story_ID=2329718 (Discusses the work of David Gil and Lera Boroditsky.)

31. Ibid.

32. Barnett, Adrian, "For Want of a Better Word," *New Scientist*, January 31, 2004. http://www.newscientist.com/opinion/ opinterview. jsp?id=ns24321 (Describes the work of linguist Alexandra Aikhenvald. She was a Research Fellow at the Institute of Oriental Studies of the Academy of Sciences of the USSR from 1980to 1989, and Professor of Linguistics at the Federal University of Santa Catarina, Brazil, until 1994.)

33. Hitt, Jack, "Say No More," *New York Times Magazine*, 52–54, February 29, 2004. (Discusses Kawesqar, the language native to Patagonia.)

34. Krattenmaker, Tom, "Swarthmore College Linguist Finds Unrecorded Language in Siberia on the Brink of Extinction," January, 21, 2004, http://www.swarthmore.edu/news/releases/04/harrison.html (On the work of K. David Harrison and the Ös.)
See also, Will Knight, "Half of All Languages Face Extinction this Century," *New Scientist*, 16 February 04, Http://www.newscientist.com/news/ news.jsp?id=ns99994685

35. Ibid.

36. Crystal, David, *The Cambridge Encyclopedia of Language* (New York: Cambridge University Press, 1987).

37. Of course, even if you and I use the same word for a color or other sensation, it doesn't mean we perceive the same thing. Imagine going to a movie theater with your date, watching the *Wizard of Oz*, and gazing at the beautiful green color of the Emerald City. Your parents taught you what green stood for. Your date's parents also taught him or her that when a particular wavelength of light intersected the retina, we should call the sensation "green." But it would be very difficult to determine if you are experiencing the same sensation as your date. In some sense, the use of the word "green" may not only shape our perceptions, or aid in our memory and categorization of reality, but it may also shield others from what we are truly perceiving. The structure of one language may suggest or even reveal relationships between concepts that may be veiled by another language.

38. Roth, Cecil, *Encyclopedia Judaica* (18 Volumes) (New York: Coronet Books Inc; reprint edition, 2002). Cecil Roth (1899–1970) was a Jewish historian and editor of *Encyclopedia Judaica* from 1965 until his death.

39. Parry, Aaron (Rabbi), *The Complete Idiot's Guide to the Talmud* (New York: Alpha, 2004), 149.

40. McKenna, Terence, *The Archaic Revival: Speculations on Psychedelic Mushrooms, the Amazon, Virtual Reality, UFOs and More* (Harper SanFrancisco, 1992), 64.

41. Morrison, Grant, "Preface." In Metzger, Richard, *Book of Lies: The Disinformation Guide to Magick and the Occult* (New York: The Disinformation Company, 2003), 9.

42. McFadden, Cynthia, Interview with Chris Langan, "ABCNews.Com: Chris Langan's IQ Sets him Apart," December 10, 2001, http://www.abcnews.go.com/onair/2020/ 2020_991210_iq_chat.html

43. Pesce, Mark, "The Executable Dreamtime." In Metzger, Richard, *Book of Lies: The Disinformation Guide to Magick and the Occult* (New York: The Disinformation Company, 2003), 9.

44. I devote an entire chapter to the brain and time in Pickover, Clifford, *Time: A Traveler's Guide* (New York: Oxford University Press, 1998). Similar information can also be found in Pickover, Clifford, *The Paradox of God* (New York: St. Martin's Press/Palgrave, 2002).

45. Pesce, Mark, "The Executable Dreamtime." In Metzger, Richard, *Book of Lies: The Disinformation Guide to Magick and the Occult* (New York: The Disinformation Company, 2003), 28.

46. Ibid., 30.

47. Here are some other strange facts for you. The followers of the ancient Greek mathematician Pythagoras were also interested in 216. Pythagoras claimed he had been both a plant and an animal in his past lives and, like Saint Francis, he preached to animals. Pythagoras and his followers believed in *anamnesis*, the recollection of one's previous incarnations. In various ancient Greek writings, we are told the exact number of years between each of Pythagoras' incarnations is 216. Interestingly, Pythagoreans considered 216 to be a mystical number, because it is 6 cubed ($6 \times 6 \times 6$). Six was also considered a "circular number" because its powers always ended in six. The fetus was considered to have been formed after 216 days.

Chapter 3
Bertrand Russell's Twenty Favorite Words

1. Clifford A. Pickover.

2. Shattuck, Roger, *Proust's Way: A Field Guide to In Search of Lost Time* (New York, W. W. Norton & Company, 2001).

3. Feinberg, Barry and Kasrils, Ronald (editors), *Dear Bertrand Russell: A Selection of His Correspondence with the General Public 1950–1968* (Boston: Houghton Mifflin, 1969).

4. Stevenson, Florence, "A History of the John C. Hart Memorial Library," http://www.wls.lib.ny.us/libs/yorktown/aboutus.htm

5. See Einstein Archive 33-155, 75-144. Also quoted in Calaprice, Alice, *The Expanded Quotable Einstein* (Princeton, New Jersey, 2000), 98.

6. Feinberg, Barry and Kasrils, Ronald (editors), *Dear Bertrand Russell: A Selection of His Correspondence with the General Public 1950–1968* (Boston: Houghton Mifflin, 1969).

7. The city of Monongahela is situated on the Monongahela River, 17 miles southeast of Pittsburgh, Pennsylvania.

8. Dickson, Paul, *Names: A Collector's Compendium of Rare and Unusual, Bold and Beautiful, Odd and Whimsical Names* (New York: Smithmark Publishers; reissue edition, 1988); Dickson, Paul, *Words* (New York: Delacorte Press, 1982); Dickson, Paul *Dickson's Word Treasury: A Connoisseur's Collection of Old and New, Weird and Wonderful, Useful and Outlandish Words* (New York: John Wiley & Sons; revised edition, 1992); Miller, Jeff, "A Collection of Word Oddities and Trivia," http://members.aol.com/gulfhigh2/words.html

9. Wallace, Irving, *The People's Almanac Presents the Book Lists No. 2* (New York: Bantam Books; reissue edition, 1981).

10. Bly, Robert, "Target: Internet," *Writer's Digest*, 84(5):30–33, May 2004. Describes Harlan Ellison's feelings about the Internet.

11. Many colleagues suggested additions to the "beauty" list. I give a listing here of what they consider as "most beautiful," in a small font, so that my publisher will not be nervous about filling pages with long lists:

Bryce Canyon, a hillside of deciduous trees in autumn, the Overture of 1812, horses, a smiling little girl, Horsehead nebula, rings of Saturn, snow-capped mountain in a forest, palm trees on a Caribbean beach, newborn human infant, looks that pass between couples who have been happily married for 30 years, Ferrari Modena, Quebec City, the late Marian Anderson's voice, Newton's solution of the brachistochrone problem, a fresh pizza Martin Luther King Junior's "I Have a Dream" speech, the feel of a hug and the sound of an "I love you" from a small child, the feeling of coming home after a long journey, a familiar smell (jasmine or honeysuckle on a summer evening breeze, baking bread) that conjures up a rich and sweet memory, human body, beehive, coral, butterfly wings, the grace of an animal moving, music of great composers, bird songs, baby's smile, valley of wildflowers, sunset, Earth from 30,000 feet, members of your preferred gender with big secondary sexual traits, *Mona Lisa*, Eiffel tower, *Guernica*, *Romeo and Juliet*, "Ode on a Grecian Urn," *The Symphonie Pathetique*, starry night with no light pollution, full moon close to horizon, deep forest at evening, "my children," ice-covered trees after a freezing rain, sunset/sunrise from an airplane, a couple on a couch drinking coffee while looking out at a snowstorm, children laughing; *Phantom of the Opera* by Andrew Lloyd Webber or *Les Miserables* by Claude-Michel Schönberg, Alain Boublil, and Herbert Kretzmer; rainbow, dawn, springtime flowers, autumn trees (not autumn leaves), "Afternoon of a Faun" by Debussy, Jaclyn Smith, the scent of lilacs, sex, the Grand Canyon, waltzes, Torvill and Dean's gold-medal performance in ice dancing in the Olympics, birds (in their better aspects), a happy baby, freedom, waves crashing against a beach, Milky Way on a dark clear night, crescent moon in twilight, a woman's face, a beautiful Sierra mountain range, a redwood forest in the great Pacific northwest, a beautiful woman (preferably smiling), a cold glass of beer, Beethoven's Op131 C minor String Quartet, Vermeer's painting of girl in blue dress with jug, pounding surf, fields of grain, livestock grazing in the pasture, the triangular patterns on a shell called *Cymbiola innexa*, Pyramids at Gizeh, Niagara Falls, the human eye, redwood trees, windswept pines on a ridge, clouds seen from above, Claudia Schiffer, mountains in autumn, Grand Tetons behind Jenny Lake, sunrise view with setting near-full moon, long-exposure color photo of Andromeda Galaxy, "cow shirr shang hu-AY yo ju-AH ga ma?", the spectrum, emeralds on black velvet, Olympic-level figure-skating pair, poetry, "Jesu, Joy of Man's Desiring" by J. S. Bach, orange/cream-cheese chocolate-chip cookie, Christie Canyon's breasts, a black night sky sprinkled with stardust, a newborn baby's laugh of joy, young female nude, soap bubble, velvet, rose, snow leopard, the "conversation" of wolves, pictures of maltese dogs or silky terriers in motion, Arabian horses or panthers running, New Jersey or San Francisco ocean waves breaking on a beach, still photographs of some dog or cat breeds in "proud full stance," a hand reached out to help, and the formulas $e^{i\pi} = -1$, $E = mc^2$, $a^2 + b^2 = c^2$, $\varepsilon_0 \oint \vec{E} \cdot d\vec{A} = \Sigma q$, and $x = (-b \pm \sqrt{b^2 - 4ac})/(2a)$.

Brad P. comments, "Both the computer chip and snow crystal are beautiful because of their complexity, order and diversity—all within a small package. Both say volumes about the beauty and intelligence of their creators."

My friend Peter A. comments, "I can answer the question for most interesting, but for beauty I need to understand the presentation: scale, context, lighting, my physical environment. Beauty is in the eye of the beholder. And even a snowflake holds no beauty in the dark."

Bets L. comments, "With the exception of ammonia, wine and a seagull's cry (and possibly sushi), these are primarily items which would be only sensed visually. I think beauty has to command a deeper call than just the eyes—to somehow engage the heart. It is not the thing that is itself beautiful, it is the feeling it calls out within us...."

Herman H. comments, "I chose the tears on a little girl because I feel that tells an entire story. Beauty is in the complexity, not just in the visual aspect, but rather in everything real or imagined that can be attached to it."

12. Goss, Michael, "Kick That Habit: Brion Gysin—His Life and Magic." In Metzger, Richard, *Book of Lies: The Disinformation Guide to Magick and the Occult* (New York: The Disinformation Company, 2003), 91. See also Woodard, David, "Brion Gysin Dream Machine," http://www.davidwoodard.com/

13. Pickover, Clifford, *The Science of Aliens* (New York: Perseus, 1999).

14. Werde, Bill, "We Got Algorithm, But How About Soul?" *New York Times*, Section 4, page 12, March 21, 2004. (Describes the Dia project to assess the most desired painting.)

15. Ibid.

16. Polti, Georges, *Thirty-Six Dramatic Situations* (New York: Kessinger Publishing, 2003).

17. Wright, Ernest, *Gadsby* (Cutchogue, New York: Buccaneer Books, 1997).

18. Ibid.

19. Forthright (a.k.a. Steve Chrisomalis), "Forthright's Phrontistery: Obscure Words and Vocabulary Resources," phrontistery.50megs.com

20. Taylor, Paul, "Oulipo," http://www.nous.org.uk/oulipo.html

21. Michael, Keith, "Near a Raven," reprinted with permission, http://users.aol.com/s6sj7gt/mikerav.htm

22. Michael, Keith, *The Anagrammed Bible*, http://users.aol.com/s6sj7gt/anabible.htm Also see Brodie, Richard and Michael Keith, *The Anagrammed Bible: Proverbs, Ecclesiastes, Song of Solomon* (Richmond, Virginia: Antan Press, 2000).

23. Raiter, Brian, "Albert Einstein's Theory of Relativity in Words of Four Letters or Less," http://www.muppetlabs.com/~breadbox/txt/al.html

24. Pickover, Clifford, *Computers and the Imagination* (New York, St. Martin's Press, 1992).

25. Rachter, *The Policeman's Beard Is Half Constructed* (New York: Warner Books, 1984).

26. Here are a few more computer-generated poems:

"A Flying Knuckle"

A flying knuckle wanders on top of the glass unicorn.
With great speed the knuckle salivates;
The unicorn grasps while smoking a vibrating cow.

"A Glowing Tongue"
A glowing tongue laughs in spite of the wavering diamond.
Very slowly the tongue shines;
The diamond shines while crushing a frigid grasshopper.

"A Quivering Bone"
A quivering bone chews while touching the golden intestine.
With a terrible shutter the bone regurgitates;
The intestine yawns close to a sexy ellipse.

"A Glistening Web"
A glistening web gyrates while touching the skinny vacuum tube.
Stubbornly the web smiles;
The vacuum tube screams above a frost-encrusted Jell-O pudding.

"A Half-Dead Avocado"
A half-dead avocado oozes while puffing the fairylike ocean.
In a frenzy the avocado grasps;
The ocean breathes while dreaming about a frigid diamond.

"A Sensuous Goose"
A sensuous goose disintegrates below the hungry ocean.
Stubbornly the goose shakes;
The ocean drools far away from a glittering knuckle.

"A Sexy Cloud"
A sexy cloud collapses while smoking the moldy soul.
Heavily the cloud frowns;
The soul evaporates behind a chocolate mouth.

"A Flawless Diamond"
A flawless diamond burns in synchrony with the lost goose.
While waving its tentacles the diamond shakes;
The goose gyrates deep within a green ellipse.

"A Religious Avocado"
A religious avocado gesticulates in between the chocolate grasshopper.
Grotesquely the avocado burns;
The grasshopper yawns while making love to a magnetic centipede.

"A Skinny Earthworm"
A skinny earthworm wriggles before the glittering brain.

> Heavily the earthworm oozes;
> The brain evolves while making love to a robotoid earthworm.

27. Stephen L. Thaler, "IEI's Revolutionary Virtual Reality Drivers for the Entertainment Industry," http://www.imagination-engines.com/papers/vrdrivers.htm

28. Ibid.

29. Donaldson, Stephen R., *The One Tree* (The Second Chronicles of Thomas Covenant, Book 2) (New York: Del Rey Books; reissue edition, 1993).

30. Artaud, Antonin, "Van Gogh, the Man Suicided by Society" (1947). Reprinted in Artaud, Antonin, *Selected Writings*, edited by Susan Sontag (Berkeley: University of California Press, 1976).

31. Pinchbeck, Daniel, *Breaking Open the Head* (New York: Broadway, 2002).

32. Swinburne, Algernon Charles, *Selected Poems* (New York: Routledge, 2002). Here is the remainder of "Amphigory":

> Nay, for the nick of the tick of the time
> is a tremulous touch on the temples of terror,
> Strained as the sinews yet strenuous with strife
> of the dead who is dumb as the dust-heaps of death:
> Surely no soul is it, sweet as the spasm
> of erotic emotional exquisite error,
> Bathed in the balms of beatified bliss,
> beatific itself by beatitude's breath.
> Surely no spirit or sense of a soul
> that was soft to the spirit and soul of our senses
> Sweetens the stress of suspiring suspicion
> that sobs in the semblance and sound of a sigh;
> Only this oracle opens Olympian,
> in mystical moods and triangular tenses—
> Life is the lust of a lamp for the light
> that is dark till the dawn of the day when we die.
> Mild is the mirk and monotonous music of memory,
> melodiously mute as it may be,
> While the hope in the heart of a hero is bruised
> by the breach of men's rapiers, resigned to the rod;
> Made meek as a mother whose bosom-beats bound
> with the bliss-bringing bulk of a balm-breathing baby,
> As they grope through the grave-yard of creeds, under skies
> growing green at a groan for the grimness of God.
> Blank is the book of his bounty beholden of old,
> and its binding is blacker than bluer:

> Out of blue into black is the scheme of the skies,
> and their dews are the wine of the bloodshed of things;
> Till the darkling desire of delight shall be free
> as a fawn that is freed from the fangs that pursue her,
> Till the heart-beats of hell shall be hushed by a hymn
> from the hunt that has harried the kennel of kings.

33. Recently, I've created Web pages that facilitate a practice I call Google Grokking. Have you ever noticed that certain word combinations lead to fascinating topics and Web pages—when entered into Google? Over the years, my readers have written to me with some of their favorite "search packs," that is, combinations of words that yield funny, profound, or mind-bending clusters of Web pages. Do you have any search packs to suggest that encourage the mind to transcend its Earthly bounds? Which of these search packs did you like best? Here are my favorites to type into Google. Type these into Google and read the pages that result.

Search 1: god synchronicity transcendence afterlife.

Search 2: mind lsd dimensions.

Search 3: third temple jerusalem.

Search 4: parallel universes beings god.

Search 5: pickover hawking jesus.

Search 6: genius iq savant.

Search 7: quiet mind tomc.

Search 8: dreams transcendence geometry.

Search 9: apocalyptic dreams time travel.

Search 10: jesus never died.

Search 11: problem miracles.

Search 12: alien inside brain.

Search 13: rephaim gilgal.

Search 14: god mathematics knowledge.

I also study "Amazon Whacking" at my Web site. Here's a challenge involving Amazon.com.

1. Start at a book listed at Amazon.com.

2. Go to another book by selecting a book on the list titled "Customers who bought this book also bought...."

3. Go to step 2.

What's the longest path you can find without repeating a book in the path? What books were in your path? How many of the books have you read and would you like to read? What can we learn from this game? In a way, this game is like following a stream through Amazon or following a rollercoaster as it whips around the track.

For example, here's a short one having a path length of 6 that leads from *The Seven Mysteries of Life* to *The Evolution of Cooperation*:

The Seven Mysteries of Life by Guy Murchie → *Butterfly Economics: A New General Theory of Social and Economic Behavior* by Paul Omerod → Paul Ormerod, *Swarm Intelligence: From Natural to Artificial Systems* → by Eric Bonabeau, Marco Dorigo → Guy Theraulaz, *Hidden Order: How Adaptation Builds Complexity* by John H. Holland → Heather Mimnaugh, *Turtles, Termites, and Traffic Jams* by Mitchel Resnick → *The Evolution of Cooperation* by Robert Axelrod.

At my Web site, people have studied paths containing chains of more than 300 books.

34. John Koch, "Interview with Stephen Pinker," *The Boston Globe Magazine*, 1999, http://pinker.wjh.harvard.edu/about/media/2000_01_23_bostonglobemagazine.html

35. Clifford Pickover, personal conversation with Joumana Medlej. Visit her Web site at www.cedarseed.com.

Chapter 4
DMT, Moses, and the Quest for Transcendence

1. Clifford A. Pickover.

2. Einstein, Albert, "On 'Cosmic Religion,' a Worship of the Harmony and Beauties of Nature That Became the Common Faith of Physicists." In Einstein, Albert, *Cosmic Religion with Other Opinions and Aphorisms* (New York: Covici-Friede, 1931), 52.

3. Armstrong, Karen, *A History of God: The 4,000-Year Quest of Judaism, Christianity and Islam* (New York: Ballantine Books; reprint edition, 1994).

4. Proust, Marcel, *In Search of Lost Time* (translated by C. K. Moncrieff and Terence Kilmartin; revised by D. J. Enright) (New York: Modern Library, 1992, 6 volumes).

5. Ibid.

6. Huxley, Aldous, *The Doors of Perception and Heaven and Hell* (New York: HarperPerennial; reissue edition, 1990).

7. Ibid.

8. Pinchbeck, Daniel, *Breaking Open the Head* (New York: Broadway, 2002).

9. McKenna, Terence, *The Archaic Revival: Speculations on Psychedelic Mushrooms, the Amazon, Virtual Reality, UFOs and More* (Harper SanFrancisco, 1992), 38.

10. Einstein, Albert, "On 'Cosmic Religion,' a Worship of the Harmony and Beauties of Nature That Became the Common faith of Physicists." In Einstein, Albert, *Cosmic Religion with Other Opinions and Aphorisms* (New York: Covici-Friede, 1931), 102.

11. Rick Strassman, personal communication.

12. Dennis McKenna, personal communication.

13. Pickover, Clifford, *The Science of Aliens* (New York: Basic Books/ Perseus, 2000).

14. Horgan, John, *Rational Mysticism* (New York: Houghton Mifflin, 2003).

15. Shanon, Benny, *The Antipodes of the Mind: Charting the Phenomenology of the Ayahuasca Experience* (New York: Oxford University Press, 2003).

16. Pinchbeck, Daniel, "Interview with Daniel Pinchbeck by Joseph Durwin, The Orbits Project," http://brainmachines.com/index2.html

17. Belgian naturalist and playwright Maurice Maeterlinck (1862–1949) published fascinating books like *The Life of the White Ant* and *The Life of the Bee* in which he describes insects working together as components of a larger superorganism. In 1911, Maeterlinck won the Nobel Prize for literature following the success of his play *The Bluebird*. In 1901, he published The Life of the Bee, a mixture of natural history and philosophy. Maeterlinck believed that insects are like alien creatures and "do not belong to our world."

18. de Alverga, Alex Polari, *O Livro das Miraçõs* (*The Book of Visions*) (Rio de Janerio: Editora Record, 1984. An English translation is now available: Polari de Alverga, Alex, *Forest of Visions: Ayahuasca, Amazonian Spirituality, and the Santo Daime Tradition* (edited and introduced by Stephen Larsen) (Rochester, Vermont: Park Street Press, 1999).

19. Montgomery, Charles, "Think LSD and Ecstasy Have Devoted Followings? The Next Drug Sliding Down the Nirvana Pipeline Has Already Spawned Three New Religions," *Vancouver Sun*, 10 February, 2001.

20. McKenna, Terence, *The Archaic Revival: Speculations on Psychedelic Mushrooms, the Amazon, Virtual Reality, UFOs and More* (HarperSanFrancisco, 1992), 64.

21. Shulgin, Alexander and Ann Shulgin, *Pihkal: A Chemical Love Story* (Berkely, California: Transform Press, 1991).

22. Meyer, Peter, "Apparent Communication with Discarnate Entities Induced by Dimethyltryptamine (DMT)," http://www.serendipity.li/dmt/dmtart00.html; see also Meyer, Peter, "Answers to DMT Interview Questions," http://www.serendipity.li/dmt/dmtinter.html

23. Westfall, Richard S., *Never At Rest: A Biography of Isaac Newton* (New York: Cambridge University Press, 1983), 351, which cites *Yahuda MS* 9.2 ff. 139.

24. William James, *The Varieties of Religious Experience*: A *Study in Human Nature* (New York: Modern Library; reprint edition, 1994); first published in1902.

25. Personal communication.

26. Barnaby, Christopher J., "Everything Is Implied: The Nature of Intraterrestrial Contact," http://www.cjbarnaby.com/aarticles_everythingisimplied.htm

27. Watts, Alan, *The Joyous Cosmology* (New York: Random House, 1962).

28. Kent, James, "The Case Against DMT Elves: James Kent Attempts to Tie a Knot in the Meme of Autonomous Elves and Other DMT Entities," http://www.tripzine.com/articles.asp?id=dmt_pickover

28. Horgan, John, *Rational Mysticism* (New York: Houghton Mifflin, 2003).

29. Ibid.

30. Ibid.

31. Laplante, Eve, *Seized: Temporal Lobe Epilepsy As a Medical, Historical, and Artistic Phenomenon* (New York: HarperCollins, 1993), now available at BackInPrint.Com.

32. Ezekiel 1: 4.

33. Laplante, Eve, *Seized: Temporal Lobe Epilepsy As a Medical, Historical, and Artistic Phenomenon* (New York: HarperCollins, 1993), now available at BackInPrint.com.

34. Jacobs, A. J., *The Know-It-All: One Man's Humble Quest to Become the Smartest Person in the World* (New York: Simon and Schuster, 2004), 9.

35. Montgomery, Charles, "Think LSD and Ecstasy Have Devoted Followings? The Next Drug Sliding Down the Nirvana Pipeline Has Already Spawned Three New Religions," *Vancouver Sun*, 10 February, 2001.

36. Gould, Stephen, quoted in Kick, Russ, *The Disinformation Book of Lists* (New York: The Disinformation Company, 2004), 17. See also Grinspoon, Lester, *Marihuana, the Forbidden Medicine* (New Haven, Connecticut: Yale University Press; revised edition, 1997), and Dana Larsen, "Stoned Scientists," *Cannabis Culture Magazine*, April 4, 2003, http://www.cannabisculture.com/articles/2783.html

Chapter 5
Brain Syndromes Open Portals to Parallel Universes

1. Clifford A. Pickover.

2. Jamison, Kay Redfield, "Manic-Depressive Illness and Creativity," *Scientific American*, 272(2): 62–67, February 1995.

3. Jamison, Kay Redfield, *An Unquiet Mind: A Memoir of Moods and Madness* (New York: Vintage, 1997).

4. Smith, Audrey, "Studies on Golden Hamsters During Cooling To and Rewarming From Body Temperatures Below 0 Degrees Centigrade," *Proceedings of the Royal Society of Biology, London Series B*. 147: 517, 1957.

5. Suda, Isamu and A. C. Kito, "Histological Cryoprotection of Rat and Rabbit Brains," *Cryoletters* 5: 33, 1966.

6. Proust, Marcel, *In Search of Lost Time: Swann's Way* (translated by C. K. Moncrieff and Terence Kilmartin; revised by D. J. Enright) (New York: Modern Library, 1992).

7. Ibid., *In Search of Lost Time: Within a Budding Grove*.

8. Doyle, Stephen J. and Maggie Harrison, "Lost in Lilliput" http://www.northerneye.co.uk/sweep.htm (Reproduced from "Health and Aging," April 1998; this article describes how Bonnet people see creatures wearing hats.)

9. Highfield, Roger, Telegraph Group Limited, "Ghosts and Witches on the Brain," http://www.telegraph.co.uk/connected/main.jhtml?xml=/connected/2002/11/06/ecfwitch06.xml (Describes the work of Dominic Ffytche, who has studied many Bonnet-people.)

10. Teunisse, Robert J., Johan R. Cruysberg, Willibrord H. Hoefnagels, André L. Verbeek, and Frans G. Zitman, "Visual Hallucinations in

Psychologically Normal People: Charles Bonnet's Syndrome," *Lancet*, 347: 794–797, March 1996. (Describes the research at the Low Vision Unit of the Department of Ophthalmology, University Hospital, Nijmegen, which shows that the Bonnet visions are sometimes comical, like "two miniature policemen guiding a midget villain.")

11. Enoch, M. David and Hadrian Ball, *Uncommon Psychiatric Syndromes* (fourth edition) (London: Arnold, 2001).

12. Pinchbeck, Daniel, *Breaking Open the Head* (New York: Broadway, 2002), 247.

13. Ramsland, Katherine, *Dean Koontz: A Writer's Biography* (New York: HarperPrism, 1997), 378.

14. Ellis, Havelock, "Mescal: A New Artificial Paradise," *The Contemporary Review*, January 1898. Henry Havelock Ellis traveled widely in Australia and South America before studying medicine in London. This text from the late 1800s describes a vision produced by mescal.

Chapter 6
From Holiday Inn to the Head of Christ

1. Clifford A. Pickover.

2. Heuct, Stephane, Marcel Proust, *Remembrance of Things Past: Combray* (graphic novel) (New York: NBM Publishing, Inc., 2002). (An excellent comic-book approach to understanding Proust for most mortals who cannot sit through 3,000 pages.)

3. Smith, Dennis, "FYI: Celery Soda?" http://www.angelfire.com/zine2/thesodafizz/2003oct11.html

4. Kirn, Walter, "Birth of a Vacation," *New York Times Magazine*, Section 6, p. 12, December 28, 2003.

5. Provine, Robert, *Laughter: A Scientific Investigation* (New York: Viking, 2000). See also, Walker, Rob, "Making Us Laugh," *New York Times Magazine*, Section 6, p. 28, December 28, 2003. (On the inventor of the laugh track.)

6. Alvy, Ted, "Cat Simril Interviews Paul Krassner," http://members.aol.com/tedalvy/cat.htm
Krassner has published material on the psychedelic revolution and has experimented with LSD with Tim Leary, Ram Dass and Ken Kesey, and later accompanied Groucho Marx on his first acid trip.

7. Thomas Jefferson, letter to Charles Thomson, January 9, 1816.

8. Thomas Jefferson, letter to Ezra Stiles Ely, June 25, 1819.

9. Thomas Jefferson, letter to John Adams, January 24, 1814.

10. Thomas Jefferson, letter to John Adams, April 11, 1823.

11. Statement published in the Romanian Jewish journal *Renasterea Noastra*, January 1933. Also published in *Mein Weltbild*; reprinted in Einstein, Albert, *Ideas and Opinions* (New York: Crown, 1954), 184–185.

12. Quoted in William Hermanns, "A Talk with Einstein," October 1943, Einstein Archive 44-285.

13. Herbert Muschamp, the *New York Times* architectural critic, wrote an interesting article about *The Arcades Project* that appeared in the Arts & Leisure section of the *New York Times* on January 16, 2000.

14. Benjamin, Walter, *Reflections: Essays, Aphorisms, Autobiographical Writings*, edited by Peter Demetz (New York: Schocken Books, 1986).

15. *Macbeth*, II.2.35–39.

16. *Macbeth*, IV.1.77.

17. Carroll, Lewis, *Through the Looking Glass: And What Alice Found There* (New York: Puffin Books; reissue edition, 1996).

18. This topic is discussed in detail in: de Botton, Alain, *How Proust Can Change Your Life* (New York, Pantheon, 1997).

19. Strauss, Neil, "He Aims! He Shoots! Yes!!" *New York Times* (Sunday Styles) Section 9, Sunday, January 25, 2004, p. 1.

20. Ibid.

21. Hannah M. G. Shapero, personal communication. Visit her Web site at www.pyracantha.com.

22. Ibid.

23. Kennedy, Randy, "Who Was That Food Stylist? Film Credits Roll On," *New York Times*, Late Edition—Final, Section 1, Page 1, Column 4, Sunday, January 11, 2004.

24. Ibid.

25. Ecclesiastes, 12:6–7.

26. de Botton, Alain, *How Proust Can Change Your Life* (New York, Pantheon, 1997).

27. Bernard, Andre, *Rotten Rejections: A Literary Companion* (Yonkers, New York: Pushcart Press, 1990).

28. Seife, Charles, "Physics Enters the Twilight Zone," *Science*, 305(5683): 464–466, July 23, 2004. Researchers suggest that if the matter and energy in the universe are created by random quantum fluctua-

tions, as cosmic inflation dictates, then there will be an infinite number of copies of the finite configuration of matter and energy in our 100-billion-light year sphere. Scientists believe that a lump of matter and energy enclosed in a finite sphere can be arranged in only a finite number of ways—due to a restriction known as the "holographic bound."

Chapter 7
The Business of Book Publishing:
Unplugged, Up Close, and Personal

1. Ancient Chinese proverb.

2. Heuet, Stephane, Marcel Proust, *Remembrance of Things Past: Combray* (graphic novel) (New York: NBM Publishing, Inc., 2002).

3. Allen, Woody, *Without Feathers* (New York: Ballantine Books; reissue edition, 1990).

4. White, Edmund, *Marcel Proust* (New York: Viking, 1999), 136.

5. Ibid.

6. Adams, Henry, *The Letters of Henry Adams: 1858–1892* (Volumes 1–3), edited by C. J. Levenson (Cambridge, Massachusetts: Harvard University Press, 1983).

7. de Balzac, Honoré, "The Pleasures and Pains of Coffee," an essay written in the 1830s, translated from the French by Robert Onopa, transcriptions of which can be found on the Web.

8. White, Edmund, *Marcel Proust* (New York: Viking, 1999), 138.

9. Scalzi, John, "Agent to the Stars: Why I Published On Line," http://www.scalzi.com/agent/impatient.html

10. Ibid.

11. Burt, Andrew, "Submitting to the Black Hole," http://www.critters.org/users/critters/blackholes/

12. Christie, Agatha, *The Floating Admiral* (New York: Jove Books; reprint edition, 1993).

13. Grossman, Mahesh, *Write a Book Without Lifting a Finger: How to Hire a Ghostwriter Even If You're On a Shoestring Budget* (New York: Finger Press, 2003).

14. White, Edmund, *Marcel Proust* (New York, Viking, 1999), 110.

15. Let's continue with some additional writing mechanics to make your novels really shine.

   a. When to Use "Like" or "As If"

The word "like" should not be used preceding a clause with a subject and a verb. Examples:

It felt like a furry ball.

It felt as if a furry ball rolled around in his stomach.

b. Split Infinitive

Don't put an adverb between "to" and "verb." Wrong: "to carefully create."

c. Wordiness

Reduce wordiness by changing: "stooped down" to "stoop," "rose up" to "rose," "penetrated through" to "penetrate," "caught sight of " to "saw," "in the event that" to "if," "at the present time " to "now," "towards" to "toward," and "besides" to "beside," and so on.

d. When to use "To Lie" and "To Lay"

The verb form of lay takes an object, and lie does not. Examples: He laid the shovel on the ground. He wanted to lie on the ground.

e. Since versus Because

"Since" should be used when time is involved. "I have been sad since the time you arrived." Use "because" when implying a cause. "I have been sad because my house burned down."

f. Each Other versus One Another

"Each other" is used when you refer to two people. "One another" is used when you refer to three or more people. Example: "Mindy and John bumped into each other."

g. Participial Phrases

Modifying phrases that start with verbs ending in "-ing" or "-ed" require a comma before the phrase. "He pushed the ball, using a can of peaches."

h. Whoever or Whomever?

If you can't determine when to use whoever or whomever, substitute the word "he." If the sentence sounds better when using "him" than "he," then use "whomever." Can you tell which of the following is correct? (1) It was as if whoever had killed them enjoyed the task. (2) It was as if whomever had killed them enjoyed the task. "It was as if he" sounds better than "it was as if him," so use whoever.

i. Further or Farther?

Farther is used to refer to physical distance. "She runs farther than I do." Further is an adverb meaning to a greater degree. "I want further training."

j. Commas and Adjectives

Separate two or more adjectives with commas if each adjective modifies the noun equally. "They are brave, studious students." Here you could replace the comma with the word "and," and the sentence would make sense. "This was a beautiful Persian carpet." (Here "beautiful" modifies the Persian carpet.)

k. Rise or Raise?

Use rise (rose, risen) when you mean to move upward. Use raise (raised) when an object is being moved upward. "Joe raised his foot." "Joe rose early in the morning."

l. On to or Onto

Use onto when you mean "to a position on." "He tossed the spider onto the table. He held on to her foot."

m. Insectlike?

Should you use "insectlike" or "insect-like?" Do not precede "like" with a hyphen unless the letter "l" would be tripled. Examples: bill-like, lifelike, businesslike, shell-like. Do precede like with a hyphen if the word is three syllables, e.g., intestine-like. Do precede like with a hyphen if the word is a proper name, e.g., Clinton-like. An exception to this rule—use Christlike. Do precede like with a hyphen if the word is a compound word. On the other hand, when "like" is a prefix, then follow with a hyphen when used as a prefix meaning "similar to," e.g., like-minded. No hyphens are used in words that have meanings of their own, e.g., likelihood, likewise, likeness.

n. Subjunctive

The subjunctive form of the verb is used to express something contrary to fact. Use "were" in all of the following: (1) If I were king... (2) I wish you were here... (3) It was as if I were...

Usually, "as if" and "as though" suggest a subjunctive mood. In the following, nothing is contrary to fact, so it is not subjunctive. "Jack didn't know what color the dog was. If the dog was black, Joe could find it in the snow."

o. Ellipses

Ellipses can be used to indicate a pause in dialogue or a trailing off of dialogue. If a complete sentence is fading, use four dots, with no space between the final word and the dots. (One of the dots serves as a period.) If a sentence fragment is trailing off, use three dots, leaving a space between the end of the final word and the first dot.

16. Maxen, Michael, "How to Manufacture a Best Seller," *New York Times Magazine*, March 1, 1998, 31–34.

17. Cannaday, Marilyn, *Bigger Than Life: The Creator of Doc Savage* (Madison, Wisconsin: University of Wisconsin Press, 1990).

18. Ramsland, Katherine, *Dean Koontz: A Writer's Biography* (New York: HarperPrism, 1997).

19. Ibid.

20. Hendrickson, Robert, *The Literary Life and Other Curiosities* (revised and expanded edition) (New York: Harcourt Brace & Company, 1994).

21. Ibid.

22. Ibid.

23. Ibid.

24. Jackson, Holbrook, *The Anatomy of Bibliomania* (Champaign, Illinois: University of Illinois Press; reprint edition, 2001).

25. Ibid. See also, Rabinowitz, Harold, and Rob Kaplan, *A Passion for Books: A Book Lover's Treasury of Stories, Essays, Humor, Love and Lists on Collecting, Reading, Borrowing, Lending, Caring for, and Appreciating Books* (New York: Three Rivers Press, 2001); Basbanes,

Nicholas, *A Gentle Madness: Bibliophiles, Bibliomanes, and the Eternal Passion for Books* (New York: Owl Books, 1999).

Chapter 8
Neoreality and the Quest for Transcendence

1. Rumi, 13th-century Sufi poet and mystic.

2. Heuet, Stephane and Marcel Proust, *Remembrance of Things Past: Combray* (graphic novel) (New York: NBM Publishing, Inc., 2002), 68.

3. Proust, Marcel, *The Sweet Cheat Gone*, Chapter 3, Volume 6 of *Remembrance of Things Past* (translated by C. K. Moncrieff and Terence Kilmartin; revised by D. J. Enright) (New York: Modern Library, 1992).

4. Harper, Charles Jr. (editor), *Spiritual Information: 100 Perspectives* (Radnor, Pennsylvania: Templeton Foundation Press, 2005).

5. "John Templeton Foundation," http://www.templeton.org

6. Davis, Erik, "Calling Cthulhu: H. P. Lovecraft's Magick Realism." In Metzger, Richard, *Book of Lies: The Disinformation Guide to Magick and the Occult* (New York: The Disinformation Company, 2003), 140.

7. Ibid., 144.

8. Kurtz, Paul, "The New Paranatural Paradigm: Claims of Communicating with the Dead," *Skeptical Inquirer*, Nov/Dec 24(6):27–31, 2000.

9. Ferris, Timothy, *The Whole Shebang* (New York: Simon & Schuster, 1997), 304. See also, Pickover, Clifford, *The Paradox of God* (New York: Saint Martin's Press/Palgrave, 2000); Pickover, Clifford, *The Stars of Heaven* (New York: Oxford University Press, 2000).

10. Ibid., 305. There is some controversy regarding just how "fine tuned" these nuclear resonances really are. For example, see Steven Weinberg, "A Designer Universe?" *The New York Review of Books*, XLVI, p. 46, October 21, 1999. Also see, Livio, M., D. Holwell, A. Weiss, and J. Truran, "On the Anthropic Significance of the Energy of the O+ Excited State of 12C at 7.644 MeV," *Nature*, July 27, 340(6231): 281, 1989. See also, Pickover, Clifford *The Stars of Heaven* (New York: Oxford University Press, 2001).

11. Ferris, Timothy, *The Whole Shebang* (New York: Simon & Schuster, 1997). Also see Hoyle, Fred, "The universe: past and present reflections," *Engineering & Science*, November 1981, 12.

12. Jastrow, Robert, "The Astronomer and God." In Varghese, Roy Abraham (editor), *The Intellectuals Speak Out About God* (Chicago: Regnery Gateway, 1984), 22.

13. The anthropic cosmological principle asserts, in part, that the laws of the universe are not arbitrary. Instead the laws are constrained by the requirement that they must permit intelligent observers to evolve. Proponents of the anthropic principle say that human existence is only possible because the constants of physics lie within certain highly restricted ranges. Physicist John Wheeler and others interpret these amazing "coincidences" as proof that human existence somehow determines the design of the universe. There are alternative explanations for why the universe appears to be fine-tuned for life, and these explanations do not require a God or designer. For example, our universe may be one among a huge number of universes. If these universes have random values for their fundamental physical constants, then, just by chance, some of them will permit life to emerge. Using this reasoning, we would be living in one of those special universes, but no designer is needed to set the parameters.

This area is of speculation is controversial. Victor J. Stenger, Emeritus Professor of Physics at the University of Hawaii, has published numerous books and articles that suggest that the conditions for the appearance of a universe with life (and heavy element nucleosynthesis) are not quite as improbable as other physicists have suggested. For more information, see his Web site http://www.colorado.edu/philosophy/vstenger/. Also see his various books listed at Amazon.com such as *The Unconscious Quantum: Metaphysics in Modern Physics and Cosmology* (Amherst, New York: Prometheus, 1995), and his paper, "Cosmythology: Is the Universe Fine-Tuned to Produce Us?" *Skeptic* 4(2), 1996. Also see the Craig-Pigliucci Debate (William Lane Craig vs. Massimo Pigliucci), "Does God Exist?" http://www.leaderu.com/offices/billcraig/docs/craig-pigliucci1.html

14. Hawking, Stephen W., *A Brief History of Time* (New York: Bantam Books, 1988), p. 123.

15. Davies, Paul, *Other Worlds* (London: Dent, 1980), 160–161, 168–169.

16. Barrow, John and Frank Tipler, *The Anthropic Cosmological Principle* (New York: Oxford University Press, 1986).

17. Davies, Paul, *The Mind of God* (New York: Simon & Schuster: 1992), 16.

18. Adams, Fred and Greg Laughlin, *The Five Ages of the Universe* (New York: The Free Press, 2000), 202–203; Smolin, Lee, *Life of the Cosmos* (New York: Oxford University Press: 1997). Roger Penrose and Stephen Hawking have suggested that the expanding universe is described by the same equations as a collapsing black hole, but with the opposite direction of time. Black holes may be the seeds for other universes. According to John Gribbin, in *Stardust* (New Haven, Connecticut: Yale University Press, 2000), the number of baby universes may be proportional to the volume of the parent universe.

19. Rucker, Rudy, *Seek!* (New York: Four Walls Eight Windows, 1999), 150–151. See the chapter "Goodbye Big Bang," which discusses Andre Linde's baby universes.

20. Pickover, Clifford, *The Loom of God* (New York: Plenum, 1997).

21. Albert Einstein to Beatrice Frohlich, December 17, 1952. Einstein Archive 59-797. Also in Calaprice, Alice, *The Expanded Quotable Einstein* (Princeton, New Jersey: Princeton University Press, 2000), 217.

22. Pickover, Clifford, *The Paradox of God and the Science of Omniscience* (New York: Saint Martin's Press/Palgrave, 2001).

Chapter 9
Oh God, Einstein's Brain and Eyes Are Missing

1. Marian Diamond, *OMNI* magazine, c. 1995. When I asked Diamond if she knew the precise date, she responded: "I have used this sentence many times in public speeches (some 800+ in all) and have no idea where it has been printed." Diamond is one of the world's great neuroscientists, a writer, and a compelling lecturer. She's Professor of Neuroanatomy at the University of California at Berkeley.

2. Calaprice, Alice, *The Expanded Quotable Einstein* (Princeton, New Jersey: Princeton University Press, 2000), 360. Also see Pickover, Clifford, *Strange Brains and Genius:The Secret Lives of Eccentric Scientists and Madmen.* (New York: Quill, 1999)

3. Brian, Denis, *Einstein: A Life* (New York: John Wiley & Sons, 1996), 437.

4. Steve Levy quoted in Wade, Nicholas, "Brain That Rocked Physics Rests in Cider Box," *Science* 201: 696, 1978.

5. Wade, Nicholas (1981), "Brain of Einstein continues peregrinations," *Science* 213: 521.

6. Straus, Ernst G., "Reminiscences." In Holton, Gerald and Yehuda Elkana (editors), *Albert Einstein: Historical and Cultural Perspectives* (Princeton, New Jersey: Princeton University Press, 1982), 417–423.

7. Brian, Denis, *Einstein: A Life* (New York: John Wiley & Sons, 1996), 438.

8. Calaprice, Alice, *The Expanded Quotable Einstein* (Princeton, New Jersey: Princeton University Press, 2000), 161.

9. Brian, Denis, *Einstein: A Life* (New York: John Wiley & Sons, 1996), 438.

10. Wallace, Jonathan, "Proust's Ruined Mirror," *The Ethical Spectacle*, 5(2), February 1999, http://www.spectacle.org/299/main.html

11. Lunacharsky, Anatoly, *On Literature and Art* (Moscow: Progress Publishers, 1965; originally written in 1922). Translator: Y. Ganuskin.

12. Kent, James, "The Ketamine Konundrum," http://www.erowid.org/chemicals/ketamine/ketamine_info3.shtml

13. Pickover, Clifford, *Time: A Traveler's Guide* (New York: Oxford University Press, 1998).

14. Proust, Marcel, *In Search of Lost Time: Swann's Way* (translated by C. K. Moncrieff and Terence Kilmartin; revised by D. J. Enright) (New York: Modern Library, 1992), 162.

15. Levenson, Thomas, "Einstein's Gift for Simplicity," *Discover*, 25(9): 48, September, 2004. (Special issue on Einstein.)

16. Ferren, Bran, "Think Tank," *Discover*, 25(9): 82, September, 2004. (Some of the world's greatest scientific minds tell us what they love—and hate—about Einstein. Special issue on Einstein.)

17. Ibid.

18. Hoggart, Simon, and Mike Hutchinson, *Bizarre Beliefs* (London: Richard Cohen Books, 1995), 34. Skeptics also suggest that this has serious implications for the many cases of "recovered memories" of child abuse. People have suffered long terms of imprisonment because of unsupported "memories" elicited by hypnosis.

19. Wambach, Helen, *Life Before Life* (New York: Bantam, 1979).

20. Snow, Chet and Helen Wambach, *Mass Dreams of the Future* (New York: McGraw Hill, 1989). Note that Chet Snow received his Ph.D. in sociology and history and taught at Columbia University before meeting psychologist Helen Wambach.

21. Paragon Online, "Prophets, Scholars and Predictors," http://binky.paragon.co.uk/features/Paranormal_ft/cttm/part2.html See also, "Star knowledge UFO conference, interview with Chet Snow," http://www.v-j-enterprises.com/skchets.html

22. Ibid.

23. Ibid.

24. *Leading Edge Newspaper*, "Interview with Chet Snow on *Mass Dreams of the Future*," http://www.leadingedgenews.com/massdreams.html

25. Ibid.

26. Goldberg, Bruce, *Past Lives, Future Lives* (New York: Ballantine, 1982).

27. Ibid., 137–139.

28. Ibid., 138.

29. Pickover, Clifford, *Dreaming the Future* (New York: Prometheus, 1998).

Chapter 10
Burning Man and the Conquest of Reality

1. Clifford A. Pickover.

2. Brown, David Jay, *Conversations on the Edge of the Apocalypse* (New York: St. Martin's/Palgrave, 2005).

3. "What is Burning Man?", http://www.burningman.com/whatisburningman/

4. Pinchbeck, Daniel, *Breaking Open the Head* (New York: Broadway, 2002).

5. Shanon, Benny, *The Antipodes of the Mind: Charting the Phenomenology of the Ayahuasca Experience* (New York: Oxford University Press, 2003).

6. Osborne, Lawrence, "Inward Bound: Stephen LaBerge Offers the Ultimate Dream Vacation, Teaching You to Control What Unfolds In Your Mind While You Sleep," *The New York Times Magazine*, Section 6: 36–39, July 18, 2004.

7. Ibid.

8. Melton, Lisa, "Dream Drug or Demon Brew?" *New Scientist*, 182(2453): 42–43, June 26, 2004.

9. Shanon, Benny, *The Antipodes of the Mind: Charting the Phenomenology of the Ayahuasca Experience* (New York: Oxford University Press, 2003).

10. Pickover, Clifford, *The Science of Aliens* (New York: Perseus, 1999). Readers should consult this book for a more complete account of alien life, alien intelligence, and related topics—together with illustrations.

11. Lamb, Annette, Larry Johnson, and Nancy Smith, "Insects," http://www.42explore.com/insects.htm

12. Phil Meyers, "Insecta," University of Michigan Museum of Zoology Animal Diversity Web, http://animaldiversity.ummz.umich.edu/site/accounts/information/Insecta.html

13. Orkin, O., "Insect Zoo—Basic Facts, Insect Numbers," Mississippi State University, http://insectzoo.msstate.edu/Students/basic.numbers.html

14. Ibid.

15. Ibid.

16. Jonas, Doris and David Jonas, *Other Senses, Other Worlds* (New York: Stein and Day, 1976). I also discuss this topic in my book *The Science of Aliens*.

17. Vines, Gail, "Psychic Birds (or What?)," *New Scientist*, 182(2453): 48–49, June 26, 2004.

18. Ibid.

19. In the early 80s, computer geniuses Danny Hillis, Brian Silverman and friends built a Tinkertoy computer that played tic-tac-toe. For those of you not familiar with Tinkertoys, they are a set of colored wooden sticks and spools with holes in them. Invented in the early 1900s, Tinkertoys are now produced by Hasbro, Inc. One of Hillis's goals was to use the computer made of Tinkertoys and fishing line as a model that suggested that any kinds of parts can be assembled to perform logical or mathematical calculations. The prototype Tinkertoy tic-tac-toe computer was made from 10,000 Tinkertoys and resided in The Boston Computer Museum for many years. (In 1999, The Boston Computer Museum closed to the public and joined forces with the Museum of Science, Boston). See Dewdney, A., *The Tinkertoy Computer and Other Machinations* (New York: W. H. Freeman & Co., 1993).

20. Louis Guenin in *Science* says, "Zygotic personhood, which does collide with embryo research, is an implausible contradiction of the Catholic church's magisterium for most of its history. Until 1869, the

church followed Aristotle's view that not until at least day 40 does an embryo develop sufficient human form to acquire an intellectual soul, that which distinguishes human from beast (*Historia Animalium* 583b). Until then, said Aquinas, 'conception is not completed.' Aristotle believed that form and matter correspond, a view known as 'hylomorphism,' from which it follows that a being without a brain cannot house an intellectual soul. Hence the wrongfulness of abortion was said to vary with time of gestation."

This quotation is from Guenin, Louis M., "Morals and Primordials," *Science*, 292(5522), 1659–1660, 1 June 2001, http://www.-sciencemag.org/cgi/content/full/292/5522/1659

21. Dvorsky, George, "Evolving Towards Telepathy: Demand for Increasingly Powerful Communications Technology Points to Our Future As a 'Techlepathic' Species," April 26, 2004 http://www. betterhumans.com/Features/Columns/Transitory_Human/column.aspx?articleID=2004-04-26-4 Dvorsky is the deputy editor of *Betterhumans* and the president of the Toronto Transhumanist Association, a nonprofit organization devoted to encouraging the use of technology to transcend limitations of the human body.

22. Farley, Peter, "The 3.5-Milliamp Joy Buzzer," *Wired*, 12(9): 45, September, 2004.

23. Lawrence, Stacy, "The New Superbugs," *Wired*, 12(9): 58, September, 2004.

24. Ibid.

25. Avasthi, Amitabh, "Can Fish Factories Make Cheap Drugs?" *New Scientist* 183(2464): 8, September, 2004.

Chapter 11
Farewell

1. Proust, Marcel, *In Search of Lost Time: Swann's Way* (translated by C. K. Moncrieff and Terence Kilmartin; revised by D. J. Enright) (New York: Modern Library, 1992).

2. White, Edmund, *Marcel Proust* (New York, Viking, 1999), 1.

3. Ibid., 3.

4. Ibid., 113.

5. Ludwig, Arnold, *The Price of Greatness: Resolving the Creativity and Madness Controversy* (New York: Guilford Press, 1995), 45. Also

see Pickover, Clifford, *Strange Brains and Genius: The Secret Lives of Eccentric Scientists and Madmen* (New York: Quill, 1999).

6. Proust, Marcel, *In Search of Lost Time: Swann's Way*, 160.

7. Ibid.

8. Ibid., 161.

9. Ibid., 163.

10. Ibid., 164.

11. Ibid., 164.

Feedback

1. The Age Company Ltd., "Give Proust a Chance," July 10, 2004, http://www.theage.com.au/articles/2004/07/08/1089000235945.html

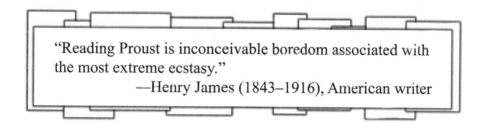

"Reading Proust is inconceivable boredom associated with the most extreme ecstasy."
—Henry James (1843–1916), American writer

Further Reading

"The more Leonardo da Vinci struggles within his chains of ignorance, the sadder it becomes....Though he breaks his fetters in many places, he never escapes from them. I wonder if in a number of fields [today], we are not in a rather similarly sad state today with the fetters being no less powerful for being unknown to us, even unfelt."

—Kenneth Keele, in Sherwin B. Nuland's
Leonardo da Vinci

General Reading

Albert Einstein

Calaprice, Alice, *The Expanded Quotable Einstein* (Princeton, New Jersey: Princeton University Press, 2000).

Brian, Denis, *Einstein: A Life* (New York: John Wiley & Sons, 1996).

Stix, Gary, "The Patent Clerk's Legacy," *Scientific American* 291(3): 44–49, September, 2004. (The entire September issue is devoted to Einstein.)

Marcel Proust

Heuet, Stephane, Marcel Proust, *Remembrance of Things Past: Combray* (graphic novel) (New York: NBM Publishing, Inc., 2002). (An excellent comic-book approach to understanding Proust for most mortals who cannot sit through 3,000 pages.)

Heuet, Stephane, Marcel Proust, *Remembrance of Things Past: Within a Budding Grove* (graphic novel) (New York: NBM Publishing, Inc., 2002).

Pinchbeck, Daniel, *Breaking Open the Head* (New York: Broadway, 2002).

Proust, Marcel, *Swann's Way* (translated by Lydia Davis; edited by Christopher Prendergast) (New York: Viking Press, 2003).

Proust, Marcel, *In Search of Lost Time*. Translated by C. K. Moncrieff and Terence Kilmartin; revised by D. J. Enright. (New York: Modern Library, 1992, 6 volumes).

Proust, Marcel, *A la recherche du temps perdu*. Texte édité et présenté par Pierre Clarac and André Ferré. Bibliothèque de la Pléiade. (Paris: Gallimard, 1954), 3 tomes.

Shattuck, Roger, *Proust's Way: A Field Guide to* In Search of Lost Time (New York, W. W. Norton & Company, 2001).

White, Edmund, *Marcel Proust* (New York, Viking, 1999)

Psychedelics
For an excellent list of great modern scientists who were inspired by marijuana and psychedelics, see: Kick, Russ, *The Disinformation Book of Lists* (New York: The Disinformation Company, 2004), 17. See also Grinspoon, Lester, *Marihuana, the Forbidden Medicine* (New Haven, Connecticut: Yale University Press; revised edition, 1997); and Dana Larsen, "Stoned Scientists," *Cannabis Culture Magazine*, April 4, 2003, http://www.cannabisculture.com/articles/2783.html

Chapters

Introduction
Pickover, Clifford, *The Paradox of God* (New York: St. Martin's Press/ Palgrave, 2002).

Chapter 1
On Fugu Sushi and Transdimensional Reality Worms
Barr, Cameron W., "Cicada: The Other, Other White Meat: Epicures Ready to Make a Meal of High-Pitched Pests," http://www.msnbc.msn.com/id/4752983/, April 16, 2004, reprints story from the *The Washington Post*. (This article discusses Grubco, Inc.)

Britton, Everard B., "A Pointer to a New Hallucinogen of Insect Origin," *Journal of Ethnopharmacology*, 12(3): 331–333, December, 1984.

Davis, Erik, "Calling Cthulhu: H. P. Lovecraft's Magick Realism." In Metzger, Richard, *Book of Lies: The Disinformation Guide to Magick and the Occult* (New York: The Disinformation Company, 2003), 138–139.

Horgan, John, *Rational Mysticism* (New York: Houghton Mifflin, 2003).

Lasswitz, Kurd. *Auf zwei Planeten* (Lepzig: Verlag B. Elischer Nachfolger, 1897). *Two Planets*, English-language edition, translated by Hans Rudruick (Carbondale, Illinois: Southern Illinois University Press, 1971.

Daniel Pinchbeck, *Breaking Open the Head* (New York: Broadway, 2002).

Samorini, Giorgio, *Animals and Psychedelics* (Rochester, Vermont: Park Street Press, 2000).

Stapledon, Olaf, *Star Maker* (Los Angeles: Tarcher; reprint edition 1987).

Chapter 2
The Quantum Mechanics of Hopi Indians

Alford, Dan Moonhaw, "Whorf Hypothesis Hoax: Sin, Suffering and Redemption in Academe," Chapter Seven from *The Secret Life of Language*, October 17, 2002 draft. http://www.enformy.com/dma-Chap7.htm "Babel's Children," *The Economist*, 370(8357): 61, January 8, 2004, http://www.economist.com/printedition/displayStory.cfm?Story_ID=2329718

Barnett, Adrian, "For Want of a Better Word," *New Scientist*, January 31, 2004. (Describes the work of linguist Alexandra Aikhenvald.) http://www.newscientist.com/opinion/opinterview.jsp?id=ns24321

Bimer, Betty (Linguistic Society of America), "Is It True That the Language I Speak Shapes My Thoughts?" http://www.lsadc.org/faq/index.php?aaa=faqthink.htm Chapoval, Tova, "World's Greatest Polyglot," from a *Reuters* wire article that appeared in the *SF Chronicle* sometime in the early 1990s, http://www.spidra.com/fazah.html (On Ziad Fazah, the world's greatest living polyglot.)

Dyson, Freeman, "The Two Windows." In Templeton, John Marks (editor), *How Large Is God*? (Radnor, Pennyslvania: Templeton Foundation Press, 1997).

Feinberg, Barry and Kasrils, Ronald (editors), *Dear Bertrand Russell: A Selection of His Correspondence with the General Public 1950–1968* (Boston: Houghton Mifflin, 1969).

Halpern, Mark, "The Eskimo Snow Vocabulary Debate: Fallacies and Confusions," http://www.rules-of-the-game.com/lin003-snow-words.htm

Harder, Ben, "Linguists in Siberia Record Dying Tongues," *Science News* 165(9): 142, February 28, 2004. (On the Ös speakers.)

Hitt, Jack, "Say No More," *New York Times Magazine,* 52–54, February 29, 2004. (Discusses Kawesqar, the language native to Patagonia.)

Jacobson, Steven A., *Yup'ik Eskimo Dictionary*. (Fairbanks, Alaska: Alaska Native Language Center, University of Alaska, Fairbanks, 1984).

Krattenmaker, Tom, "Swarthmore College Linguist Finds Unrecorded Language in Siberia on the Brink of Extinction," January, 21, 2004, http://www.swarthmore.edu/news/releases/04/harrison.html (On the work of K. David Harrison and the Ös.) See also, Knight, Will, "Half of All Languages Face Extinction This Century," *New Scientist,* 16 February 04, http://www.newscientist.com/news/news.jsp?id=ns99994685

"Linguistic Relativism and Korean," http://www.emptybottle.org/glass/2003/04/linguistic_relativism_and_korean.php

Magat, Rabbi Dana, "Welcome to Temple Emanu-el," http://www.templesanjose.org/JudaismInfo/ (Provided information on the names of God.)

Medlej, Joumana, "Knock, Knock," http://www.cedarseed.com/ And in particular, see "Thinking in Tongues," http://www.cedarseed.com/air/blabla3.html

McFadden, Cynthia, Interview with Chris Langan, "ABCNews.Com: Chris Langan's IQ Sets him Apart," December 10, 2001, http://www.abcnews.go.com/onair/2020/2020_991210_iq_chat.html

Morrison, Grant, "Preface." In Metzger, Richard, *Book of Lies: The Disinformation Guide to Magick and the Occult* (New York: The Disinformation Company, 2003), 9.

Nunberg, Geoffrey. "Snowblind," *Natural Language and Linguistic Theory* 14: 205–213, February, 1996.

Pesce, Mark, "The Executable Dreamtime." In Metzger, Richard, *Book of Lies: The Disinformation Guide to Magick and the Occult* (New York: The Disinformation Company, 2003), 9.

Pinker, Steve, *The Language Instinct (New York: Harper*Perennial, 2000).

Pinker, Steve, *The Language Instinct,* http://www.ripon.edu/academics/global/languageinstinct.html

Pullum, Geoffrey. *The Great Eskimo Vocabulary Hoax and Other Irreverent Essays on the Study of Language* (Chicago: University of Chicago Press, 1991).

Raiter, Brian, "Albert Einstein's Theory of Relativity in Words of Four Letters or Less," http://www.muppetlabs.com/~breadbox/txt/al.html

Ross, Philip, "Draining the Language Out of Color," *Scientific American*, 290(4): 46–47, April 2004. (Discusses the work of linguist Paul Kay.)

Thaler, Stephen, "Imagination Engines, Inc." http://www.imagination-engines.com/

Whorf, Benjamin, *Language, Thought and Reality: Selected Writings of Benjamin Lee Whorf* (edited by John B. Carroll) (Cambridge, Massachusetts: MIT Press, 1964).

Woodbury, Anthony C., "Counting Eskimo Words for Snow: A Citizen's Guide," University of Texas at Austin, July 1991, *LINGUIST List*: Volume 5-1239. (Discusses lexemes referring to snow and snow-related notions in Steven A. Jacobson's *Yup'ik Eskimo Dictionary*.) http://www.princeton.edu/~browning/snow.html Note that Professor Anthony Woodbury's Web site at the Department of Linguistics, University of Texas is http://www.utexas.edu/cola/depts/linguistics/

Chapter 3
Bertrand Russell's Twenty Favorite Words

Bly, Robert, "Target: Internet," *Writer's Digest*, 84(5):30–33, May 2004. (Describes Harlan Ellison's feelings about the Internet.)

Feinberg, Barry and Kasrils, Ronald (editors), *Dear Bertrand Russell: A Selection of His Correspondence with the General Public 1950–1968* (Boston: Houghton Mifflin, 1969). (This contains his twenty favorite words.)

Goss, Michael, "Kick that Habit: Brion Gysin—His Life and Magic." In Metzger, Richard, *Book of Lies: The Disinformation Guide to Magick and the Occult* (New York: The Disinformation Company, 2003), 91.

Kurzweil, Ray, *The Age of Intelligent Machines*. (Cambridge, Massachusetts: MIT Press, 1990). (This book contains information on pattern recognition, the science of art, computer-generated poetry, and artificial intelligence.)

McKenna, Terence, *The Archaic Revival: Speculations on Psychedelic Mushrooms, the Amazon, Virtual Reality, UFOs and More* (HarperSanFrancisco, 1992), 64.

Pickover, Clifford, *The Science of Aliens* (New York: Basic Books/Perseus, 2000).

Reichardt, Jasia, *Cybernetic Serendipity: The Computer and the Arts* (New York: Praeger, 1969). (This book has a chapter on Japanese haiku, computer texts, high-entropy essays, fairy tales, and fake physics essays.)

Shattuck, Roger, *Proust's Way: A Field Guide to In Search of Lost Time* (New York: W. W. Norton & Company, 2001).

Werde, Bill, "We Got Algorithm, But How About Soul?" *New York Times*, Section 4, page 12, March 21, 2004. (Describes the Dia project to assess the most desired painting.)

Woodard, David, "Brion Gysin Dream Machine," http://www.davidwoodard.com/

Chapter 4
DMT, Moses, and the Quest For Transcendence

Bower, Bruce, "Blocked Gene Gives Mice Super Smell," *Science News* 165(9): 141, February 28, 2004.

de Alverga, Alex Polari, *O Livro das Mirações* (*The Book of Visions*) (Rio de Janerio: Editora Record, 1984). An English translation is now available: Alex Polari de Alverga, *Forest of Visions: Ayahuasca, Amazonian Spirituality, and the Santo Daime Tradition* (edited and introduced by Stephen Larsen) (Rochester, Vermont: Park Street Press, 1999).

Hayes, Charles, "Is Taking a Psychedelic an Act of Sedition?" *Tikkun*, March/April 2002, http://www.tikkun.org/magazine/index.cfm/action/tikkun/issue/tik0203/article/020313c.html. *Tikkun: A Bimonthly Jewish and Interfaith Critique of Politics, Culture, & Society* is a print magazine published in Berkeley, edited by Rabbi Michael Lerner.

Hester, Jan, "Mantis Creatures Join Alien Troops," *UFO Magazine and Phenomena Report*, 12(4): 25, July/August, 1997. (Discusses the praying mantis in alien abductions.)

Horgan, John, *Rational Mysticism* (New York: Houghton Mifflin, 2003).

Jamison, Kay, "Manic-Depressive Illness and Creativity," *Scientific American*, February. 272(2): 62–67, 1995.

Kottmeyer, Martin, "Culture Swarms with Alien Insects: Bugs Baroque," *UFO Magazine and Phenomena Report*, 12(4): 20–24, July/August, 1997. (Discusses insects in literature and in alien abductions.)

Kottmeyer, Martin, "Graying Mantis," *The Rational Examination Association of Lincoln Land (REALL) News*, http://www.reall.org/

newsletter/v07/n05/graying-mantis.html (Discusses Joe Lewels and the praying mantis God of Moses.)

Laplante, Eve, *Seized: Temporal Lobe Epilepsy As a Medical, Historical, and Artistic Phenomenon* (New York: HarperCollins, 1993); now available at BackInPrint.com.

Manuel, Frank, E., *The Religion of Isaac Newton* (Oxford: Oxford University Press, 1974).

McKenna, Terence, *The Archaic Revival: Speculations on Psychedelic Mushrooms, the Amazon, Virtual Reality, UFOs and More* (HarperSanFrancisco, 1992), 62.

McKenna, Terence, *Food of the Gods: The Search for the Original Tree of Knowledge, A Radical History of Plants, Drugs, and Human Evolution* (New York: Bantam, 1993).

McKenna, Terence, *True Hallucinations: Being an Account of the Author's Extraordinary Adventures in the Devil's Paradise* (San Francisco: HarperSanFrancisco; reprint edition, 1994).

Montgomery, Charles, "Think LSD and Ecstasy Have Devoted Followings? The Next Drug Sliding Down the Nirvana Pipeline Has Already Spawned Three New Religions," *Vancouver Sun*, 10 February 2001.

Shanon, Benny, *The Antipodes of the Mind: Charting the Phenomenology of the Ayahuasca Experience* (New York: Oxford University Press, 2003).

Strassman, Rick, M.D., *DMT: The Spirit Molecule: A Doctor's Revolutionary Research into the Biology of Near-Death and Mystical Experiences* (Rochester, Vermont: Inner Traditions International Limited, 2001).

Mack, John, M.D., *Abduction* (New York: Ballantine; revised edition, 1995).

Meyer, Peter, "Answers to DMT Interview Questions," http://www.serendipity.li/dmt/dmtinter.html

Meyer, Peter, "Apparent Communication with Discarnate Entities Induced by Dimethyltryptamine (DMT)," http://www.serendipity.li/dmt/dmtart00.html

Williams, Kevin, "City of Light: Near Death Experiences," http://www.near-death.com/experiences/research19.html

Strassman, Rick, M.D., "The Pineal Gland: Current Evidence for Its Role in Consciousness," *Psychedelic Monographs and Essays*, Volume 5 (1991), 167–205.

Pickover, Clifford, *Strange Brains and Genius: The Secret Lives of Eccentric Scientists and Madmen* (New York: Quill, 1999).

Pickover, Clifford, *The Loom of God* (New York: Plenum, 1997).

Pinchbeck, Daniel, *Breaking Open the Head* (New York: Broadway, 2002), 240.

Pinchbeck, Daniel, "Interview with Daniel Pinchbeck by Joseph Durwin, The Orbits Project," http://brainmachines.com/index2.html

Strieber, Whitley, *Communion* (New York: Avon, 1987).

Chapter 5
Brain Syndromes Open Portals to Parallel Universes

Doyle, Stephen J. and Maggie Harrison, "Lost in Lilliput," http://www.northerneye.co.uk/sweep.htm (Reproduced from "Health and Aging," April 1998; this article describes how Bonnet people see creatures wearing hats.)

Enersen, Ole Daniel, "Capgras Syndrome," http://www.whonamedit.com/synd.cfm/2535.html

Enoch, M. David and Hadrian Ball, *Uncommon Psychiatric Syndromes* (fourth edition) (London: Arnold, 2001).

Hayes, Charles, "Is Taking a Psychedelic an Act of Sedition?" *Tikkun*, March/April 2002, http://www.tikkun.org/magazine/index.cfm/action/tikkun/issue/tik0203/article/020313c.html

Highfield, Roger, Telegraph Group Limited, "Ghosts and witches on the brain," http://www.telegraph.co.uk/connected/main.jhtml?xml=/connected/2002/11/06/ecfwitch06.xml (Describes the work of Dominic Ffytche who has studied many Bonnet people.)

Hoffmann, Karen, "Victims of Capgras syndrome often cannot recognize their own image," *Post-Gazette*, Pittsburgh, Pennsylvania, Tuesday, September 23, 2003, http://www.post-gazette.com/pg/03266/224822.stm

Jamison, Kay Redfield, "Manic-Depressive Illness and Creativity," *Scientific American*, 272(2): 62–67, February 1995.

Jamison, Kay Redfield, *Touched with Fire: Manic-Depressive Illness and the Artistic Temperament* (New York: Free Press/Macmillan, 1993).

Jamison, Kay Redfield, *An Unquiet Mind: A Memoir of Moods and Madness* (New York: Vintage, 1997).

Royal National Institute of the Blind, "Charles Bonnet Syndrome," http://www.rnib.org.uk/xpedio/groups/public/documents/publicwebsite/ public_rnib003641.hcsp (Describes how Bonnet people also sometimes see vortices or entire landscapes.)

Smith, Audrey, "Studies on golden hamsters during cooling to and rewarming from body temperatures below 0 degrees centigrade," *Proceedings of the Royal Society of Biology, London Series B*. 147: 517, 1957.

Suda, I., and A. C. Kito, "Histological cryoprotection of rat and rabbit brains," *Cryoletters* 5: 33, 1966.

Teunisse, Robert J., Johan R. Cruysberg, Willibrord H. Hoefnagels, André L. Verbeek, and Frans G. Zitman, "Visual hallucinations in psychologically normal people: Charles Bonnet's yndrome," *Lancet*, 347: 794–97, March 1996. (Describes the researcher at the Low Vision Unit of the Department of Ophthalmology, University Hospital, Nijmegen, which shows that the Bonnet visions are sometimes comical, like "two miniature policemen guiding a midget villain.")

Chapter 6
From Holiday Inn to the Head of Christ

Benjamin, Walter, *The Arcades Project* (Cambridge, Massachusetts: Belknap Press, 1999).

Buck-Morss, Susan, *The Dialectics of Seeing: Walter Benjamin and the Arcades Project* (Cambridge, Massachusetts: MIT Press, 1991).

Chown, Marcus, *The Universe Next Door: The Making of Tomorrow's Science* (New York: Oxford University Press, 2002).

de Botton, Alain, *How Proust Can Change Your Life* (New York: Pantheon, 1997).

"Definition of Elf—wordIQ Dictionary & Encyclopedia" http:// www.wordiq.com/definition/Elf, see also, "Elf, from Wikipedia, the Free Encyclopedia," http://en.wikipedia.org/wiki/Elf

Elkins, James, *Why Are Our Pictures Puzzles? On the Modern Origins of Pictorial Complexity* (New York: Routledge, 1999).

Glenn, II, Ben, "A Short History of the Laugh Track," http:// www.tvparty.com/laugh.html

"Glowing Alcopops," *New Scientist*, April 10, 2004, 182(2442): 24. (Information on Don Keiller's glowing drinks.)

Gosline, Anna, "Creative Spark Can Come from Schizophrenia," *New Scientist*, 183(2457): 14, July 24, 2004.

Hammond, L. James, "Selections from Proust," http://www.ljhammond.com/proust.htm

Kennedy, Randy, "Who Was That Food Stylist? Film Credits Roll On," *New York Times*, Late Edition—Final , Section 1, Page 1, Column 4, Sunday, January 11, 2004.

Kirn, Walter, "Birth of a Vacation," *New York Times Magazine*, Section 6, p. 12, December 28, 2003.

Krassner, Paul, "Paul Krassner Biography," http://www.paul-krassner.com/pkbio.htm

Legon, Jeordan, "From science and computers, a new face of Jesus," CNN, Thursday, December 26, 2002, http://www.cnn.com/2002/TECH/science/12/25/face.jesus/

Lemley, Brad, "A Tangled Life," *Discover*, 25(9): 30, September, 2004. (Special issue on Einstein. This article mentions Einstein's schizophrenic son.)

Morgan, David, *Icons of American Protestantism: The Art of Warner Sallman* (New Haven, Connecticut: Yale University Press, 1996).

Nutt, Alfred, "The Fairy Mythology of Shakespeare," http://www.geocities.com/TimesSquare/Metro/6804/LilladarianShakespeare.html

Pickover, Clifford, *The Paradox of God* (New York: St. Martin's Press/Palgrave, 2002).

"Quantum Immortality, from Wikipedia, the Free Encyclopedia," http://en.wikipedia.org/wiki/Quantum_immortality

Shloss, Carol Loeb, *Lucia Joyce: To Dance in the Wake* (New York: Farrar Straus & Giroux, 2003).

Smith, Dennis, "FYI: Celery Soda?" http://www.angelfire.com/zine2/thesodafizz/2003oct11.html

Strauss, Neil, "He Aims! He Shoots! Yes!!" *New York Times* (Sunday Styles) Section 9, Sunday, January 25, 2004, p. 1.

Walker, Rob, "Making Us Laugh," *New York Times Magazine*, Section 6, p. 28, December 28, 2003. (On the inventor of the laugh track.)

Williams, Kevin, "The NDE and the Silver Cord: Kevin Williams' Research Conclusions," http://www.near-death.com/experiences/research12.html

Zahn, Paula, and Fillion, Mike, on CNN American Morning with Paula Zahn: Interview with Mike Fillon, writer of "The Real Face of Jesus," aired December 25, 2002—08:33 ET, http://www.cnn.com/TRANSCRIPTS/0212/25/ltm.07.html

Chapter 7
The Business of Book Publishing: Unplugged, Up Close, and Personal

Grossman, Mahesh, *Write a Book Without Lifting a Finger: How to Hire a Ghostwriter Even if You're on a Shoestring Budget* (New York: Finger Press, 2003).

Hendrickson, Robert, *The Literary Life and Other Curiosities* (revised and expanded edition) (New York: Harcourt Brace & Company, 1994).

Ramsland, Katherine, *Dean Koontz: A Writer's Biography* (New York: HarperPrism, 1997).

Scalzi, John, "Agent to the Stars," http://www.scalzi.com/agent/impatient.html

White, Edmund, *Marcel Proust* (New York, Viking, 1999), 110.

Chapter 8
Neoreality and the Quest for Transcendence

Proust, Marcel (author), Andreas Mayor (translator), Terence Kilmartin (translator), *In Search of Lost Time Volume VI, Time Regained* (Modern Library, 1999). *Time Regained*, the final volume of *In Search of Lost Time*, begins in the bleak and uncertain years of World War I. See also: Gilles Taurand, "Chicago Reader Movie Review: *The Sweet Cheat, Time Regained,* Directed by Raul Ruiz," http://www.chireader.com/movies/archives/2000/0700/000721.html

Pickover, Clifford, *Liquid Earth* (Lighthouse Point, Florida: Lighthouse Press, 2002).

Pickover, Clifford, *The Lobotomy Club* (Lighthouse Point, Florida: Lighthouse Press, 2002).

Pickover, Clifford, *Sushi Never Sleeps* (Lighthouse Point, Florida: Lighthouse Press, 2002).

Pickover, Clifford, *Egg Drop Soup* (Lighthouse Point, Florida: Lighthouse Press, 2002).

Pickover, Clifford, "Neoreality and the Quest for Transcendence." In Harper, Charles L. Jr. (editor), *Spiritual Information: 100 Perspectives* (Radnor, Pennsylvania: Templeton Foundation Press, 2005).

Pickover, Clifford, *The Stars of Heaven* (New York, Oxford University Press, 2001).

Pickover, Clifford, *The Paradox of God and the Science of Omniscience* (New York, Oxford University Press, 2001).

Chapter 9
Oh God, Einstein's Brain and Eyes Are Missing

Adams, Cecil, *The Straight Dope* (New York: Ballantine, 1988).

Brian, Denis, *Einstein: A Life* (New York: John Wiley & Sons, 1996).

Calaprice, Alice, *The Expanded Quotable Einstein* (Princeton, New Jersey: Princeton University Press, 2000).

Diamond, Marian C., Arnold B. Scheibel, Greer M. Murphy and Thomas Harvey, "On the Brain of a Scientist: Albert Einstein," *Experimental Neurology*. 88: 198–204, 1985.

Hines, Terence, "Further on Einstein's Brain," *Experimental Neurology*, 150:343–344, 1998. This short paper discusses some of the possible problems that Hines sees with the Einstein brain paper published in 1985. See also, Hines, Terence, "Einstein's Brain," *Scientific American*, 291(2): 12, August, 2004.

Kantha, S., "Albert Einstein's dyslexia and the significance of Brodmann Area 39 of his left cerebral cortex," *Medical Hypotheses* 37: 119–122, 1992.

Levenson, Thomas, "Einstein's Gift for Simplicity," *Discover*, 25(9): 48, September, 2004. (Special issue on Einstein.)

Pickover, Clifford, *Time: A Traveler's Guide* (New York: Oxford University Press, 1998).

Pickover, Clifford, *Dreaming the Future* (New York: Prometheus, 1998).

Pickover, Clifford, *Strange Brains and Genius* (New York: Quill, 1999).

Ramond, Pierre, "Think Tank," *Discover*, 25(9): 78, September, 2004. (Some of the world's greatest scientific minds tell us what they love—and hate—about Einstein. Special issue on Einstein.)

Roboz-Einstein, Elizabeth, *Hans Albert Einstein: Reminiscences of His Life and Our Life Together* (Iowa City: Iowa Institute of Hydraulic Research, 1991).

Stix, Gary, "The Patent Clerk's Legacy," *Scientific American* 291(3): 44–49, September, 2004.

Straus, Ernst G., "Reminiscences." In Holton, Gerald and Elkana, Yehuda (editors), *Albert Einstein: Historical and Cultural Perspectives* (Princeton, New Jersey: Princeton University Press, 1982), 417–423.

Tryon, Edward, "Think Tank," *Discover*, 25(9): 77, September, 2004. (Some of the world's greatest scientific minds tell us what they love—and hate—about Einstein. Special issue on Einstein.)

Wade, Nicholas, "Brain that Rocked Physics Rests in Cider Box," *Science* 201: 696, 1978.

Wade, Nicholas, "Brain of Einstein Continues Peregrinations," *Science* 213: 521, 1981.

Chapter 10
Burning Man and the Conquest of Reality

Brockman, John, *The Greatest Inventions of the Past 2,000 Years* (New York: Simon & Schuster; January 12, 2000).

Brown, David Jay and Rebecca McClen Novick, *Mavericks of the Mind: Conversations for the New Millennium* (Freedom, California: Crossing Press, 1993). (To find out more about David's work, visit his award-winning Web site: www.mavericksofthemind.com.)

Brown, David Jay, *Conversations on the Edge of the Apocalypse* (New York: Palgrave/St. Martin's, 2005).

Brown, David Jay and Rebecca McClen Novick, *Voices from the Edge: Conversations with Jerry Garcia, Ram Dass, Annie Sprinkle, Matthew Fox, Jaron Lanier, & Others* (Freedom, California: Crossing Press, 2000).

Dyer, Geoff, *Yoga for People Who Can't Be Bothered to Do It* (New York: Vintage Books, 2003).

Pickover, Clifford, *The Loom of God* (New York: Plenum, 1998).

Leonardo's Special Issue on "The Art of Burning Man," Volume 36, Number 5, 2003 (Cambridge, Massachusetts: MIT Press). I've been privilege to sit on the *Leonardo* Editorial Board for a number of years and am continually fascinated by the range and diversity of topics covered in this impressive journal. *Leonardo* began international publication in 1968, and continues to focus on writings by artists who work with technology-based art media. Most of the artists use scientific tools, and the journal serves art and science communities by promoting work at the intersection of the arts, sciences, and technology. This issue is devoted to the art of Burning Man and includes articles and photos edited by Burning Man documentary writer Louis Brill and art curator LadyBee.

Farewell

Ludwig, Arnold, *The Price of Greatness: Resolving the Creativity and Madness Controversy* (New York: Guilford Press, 1995).

Pickover, Clifford, *Strange Brains and Genius* (New York: Quill, 1999).

Proust, Marcel (author), Andreas Mayor (translator), Terence Kilmartin (translator), *In Search of Lost Time Volume VI, Time Regained* (New York: Modern Library, 1999). *Time Regained*, the final volume of *In Search of Lost Time*, begins in the bleak and uncertain years of World War I. See also: Gilles Taurand, "Chicago Reader Movie Review: The Sweet Cheat, *Time Regained*, Directed by Raul Ruiz," http://www.chireader.com/movies/archives/2000/0700/000721.html

Proust, Marcel, *Swann's Way* (New York, Penguin, 1957) reprints C. K. Scott Moncrieff's original translation of *Du côté de chez Swann*, the first of seven volumes of *A la recherche du temps perdu,* which was published in 1922, two months before Proust died.

White, Edmund, *Marcel Proust* (New York, Viking, 1999), 110.

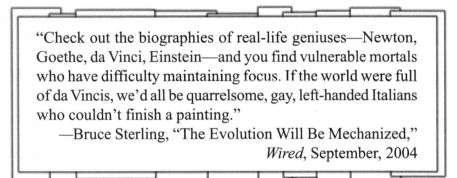

"Check out the biographies of real-life geniuses—Newton, Goethe, da Vinci, Einstein—and you find vulnerable mortals who have difficulty maintaining focus. If the world were full of da Vincis, we'd all be quarrelsome, gay, left-handed Italians who couldn't finish a painting."

—Bruce Sterling, "The Evolution Will Be Mechanized," *Wired*, September, 2004

Index

About the Author

Clifford A. Pickover received his Ph.D. from Yale University's Department of Molecular Biophysics and Biochemistry. He graduated first in his class from Franklin & Marshall College, after completing the four-year undergraduate program in three years. His many books have been translated into Italian, French, Greek, German, Japanese, Chinese, Korean, Portuguese, Spanish, Turkish, and Polish. One of the most prolific and eclectic authors of our time, Pickover is author of the popular books: *A Passion for Mathematics* (Wiley, 2005), *Calculus and Pizza* (Wiley, 2003), *The Paradox of God and the Science of Omniscience* (Palgrave/St. Martin's Press, 2002), *The Stars of Heaven* (Oxford University Press, 2001), *The Zen of Magic Squares, Circles, and Stars* (Princeton University Press, 2001), *Dreaming the Future* (Prometheus, 2001), *Wonders of Numbers* (Oxford University Press, 2000), *The Girl Who Gave Birth to Rabbits* (Prometheus, 2000), *Surfing Through Hyperspace* (Oxford University Press, 1999), *The Science of Aliens* (Basic Books, 1998), *Time: A Traveler's Guide* (Oxford University Press, 1998), *Strange Brains and Genius: The Secret Lives of Eccentric Scientists and Madmen* (Plenum, 1998), *The Alien IQ Test* (Basic Books, 1997), *The Loom of God* (Plenum, 1997), *Black Holes—A Traveler's Guide* (Wiley, 1996), and *Keys to Infinity* (Wiley, 1995). He is also author of numerous other highly acclaimed books including *Chaos in Wonderland: Visual Adventures in a Fractal World* (1994), *Mazes for the Mind: Computers and the Unexpected:* (1992), *Computers and the Imagination* (1991), and *Computers, Pattern, Chaos and Beauty* (1990), all published by St. Martin's Press—as well as the author of over 200 articles concerning topics in science, art, and mathematics. He is also coauthor, with Piers Anthony, of *Spider Legs*, a science-fiction novel once listed as Barnes and Noble's second best-selling science-fiction title. Pickover is currently an associate editor for the scientific journal *Computers and Graphics* and is an editorial board member for *Odyssey, Leonardo,* and *YLEM*.

Editor of the books *Chaos and Fractals: A Computer Graphical Journey* (Elsevier, 1998), *The Pattern Book: Fractals, Art, and Nature*

(World Scientific, 1995), *Visions of the Future: Art, Technology, and Computing in the Next Century* (St. Martin's Press, 1993), *Future Health* (St. Martin's Press, 1995), *Fractal Horizons* (St. Martin's Press, 1996), and *Visualizing Biological Information* (World Scientific, 1995), and coeditor of the books *Spiral Symmetry* (World Scientific, 1992) and *Frontiers in Scientific Visualization* (Wiley, 1994), Dr. Pickover's primary interest is finding new ways to continually expand creativity by melding art, science, mathematics, and other seemingly disparate areas of human endeavor. Pickover is also the author of the popular "Neoreality" science-fiction series (*Liquid Earth, Sushi Never Sleeps, The Lobotomy Clu*b, and *Egg Drop Soup*) in which characters explore strange realities.

The *Los Angeles Times* recently proclaimed, "Pickover has published nearly a book a year in which he stretches the limits of computers, art and thought." Pickover received first prize in the Institute of Physics' "Beauty of Physics Photographic Competition." His computer graphics have been featured on the cover of many popular magazines, and his research has recently received considerable attention by the press—including CNN's "Science and Technology Week," The Discovery Channel, *Science News*, *The Washington Post*, *Wired*, and *The Christian Science Monitor*—and also in international exhibitions and museums. *OMNI* magazine described him as "Van Leeuwenhoek's twentieth century equivalent." *Scientific American* several times featured his graphic work, calling it "strange and beautiful, stunningly realistic." *Wired* magazine wrote, "Bucky Fuller thought big, Arthur C. Clarke thinks big, but Cliff Pickover outdoes them both." Pickover holds over thirty U.S. patents, mostly concerned with novel features for computers.

For many years, Dr. Pickover was the lead columnist for *Discover* magazine's Brain-Boggler column, and he currently writes the Brain-Strain column for *Odyssey*. His puzzle calendars and cards are designed for both children and adults and have sold hundreds of thousands of copies.

Dr. Pickover's hobbies include the practice of Ch'ang-Shih Tai-Chi Ch'uan and Shaolin Kung Fu, raising golden and green severums (large Amazonian fish), and piano playing (mostly jazz). He is also a member of The SETI League, a group of signal processing enthusiasts who systematically search the sky for intelligent, extraterrestrial life. Visit his Web site, which has received over a million visits: http://www.pickover.com. He can be reached at P.O. Box 549, Millwood, New York 10546-0549 USA.

"Waiting for Time Travelers" Experiment

Given current trends in scientific progress, someday it may be possible for beings in the future to send a message to the past. Similarly, if advanced beings exist in parallel universes, then perhaps they are already trying to send us a message. In either case, their technologies should be sufficiently advanced for them to insert a message inside this box:

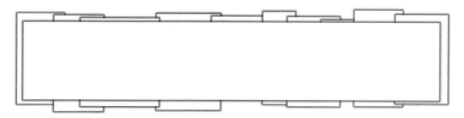

We await the appearance of a message.